International Seminar on Interdisciplinary Problems in Additive Technologies 2015

IOP Conference Series: Materials Science and Engineering
Volume 140

Tomsk, Russia
16 - 17 December 2015

ISBN: 978-1-5108-2819-3
ISSN: 1757-8981

Printed from e-media with permission by:

Curran Associates, Inc.
57 Morehouse Lane
Red Hook, NY 12571

Some format issues inherent in the e-media version may also appear in this print version.

Printed by Curran Associates, Inc. (2016)

For permission requests, please contact the Institute of Physics
at the address below.

Institute of Physics
Dirac House, Temple Back
Bristol BS1 6BE UK

Phone: 44 1 17 929 7481
Fax: 44 1 17 920 0979

techtracking@iop.org

Additional copies of this publication are available from:

Curran Associates, Inc.
57 Morehouse Lane
Red Hook, NY 12571 USA
Phone: 845-758-0400
Fax: 845-758-2633
Email: curran@proceedings.com
Web: www.proceedings.com

Table of contents

Volume 140

International Seminar on Interdisciplinary Problems in Additive Technologies
16–17 December 2015, Tomsk, Russia

Accepted papers received: 8 July 2016
Published online: 8 August 2016

Preface

Papers

IOP Conf. Series: Materials Science and Engineering **140** (2016) 011001 doi:10.1088/1757-899X/140/1/011001

Preface

Additive manufacturing is an extremely wide term and covers many conventional technologies. They include the creation of semiconductor heterostructures by methods of molecular beam epitaxy and methods of coating deposition from gas phase, offset lithography and application of multilayer coatings on products using different techniques, etc. Additive manufacturing is the process of making objects from a 3D-model through layer-by-layer binding of materials, as opposed to subtractive manufacturing methods such as traditional machining. These are fast developing technologies in various sectors, from space to medicine.

Modern additive manufacturing is rooted in stereolithography, which expanded from the idea of Otto John Munz in 1951 and advanced further due to the development of model scanning methods, techniques for forming layers from various materials, projection of images and curing technologies. Laminated object manufacturing was developed in 1985, and fused deposition modeling in 1986. Beginning from the 1990s, additive technologies have been used to produce metallic objects through controlled heating by laser or electron beams.

From 1995, a short term method for creating three-dimensional objects was born—3D printing. The surge in the development of additive technologies (AT) was in 2009. However, as usual, the energetic start was replaced by particular disappointment and the manifestation of unsuspecting drawbacks of additive technologies. Significant problems are connected with materials science: materials and products produced by AT usually do not possess the required properties and often have a limited lifetime. AT theoretically allow the production of parts of any shape and complexity. However, the products from polymers made by stereolithography do not exhibit long-term stability. The quality of the surface of a product made by the layer-by-layer cladding method is far behind that obtained by molding. Products made by jet printing are very brittle. The technologies of laser melting hardly allow complete melting of all powder particles in a layer, especially in the case of alloyed powder. So the discussion of problems in materials science, typical for various types of additive manufacturing, is an important stage in the progress of the field.

All of these problems were on the agenda of the International Seminar on Interdisciplinary Problems in Additive Technologies held on December 16–17, 2015 in Tomsk, Russia. The following topics were discussed in the seminar:

1. Materials synthesis in combined technologies;
2. Behavior, testing and control of new materials;
3. Applications of additive technologies;
4. Materials for additive technologies;

5. Mathematical modeling of new material synthesis in additive technologies, conjugate and coupling problems for modern manufacturing technologies.

The works in the proceedings cover the scope of the discussed problems.

Peer review statement

All papers published in this volume of *IOP Conference Series: Materials Science and Engineering* have been peer reviewed through processes administered by the proceedings Editors. Reviews were conducted by expert referees to the professional and scientific standards expected of a proceedings journal published by IOP Publishing.

Forming a single layer of a composite powder based on the Ti-Nb system via selective laser melting (SLM)

A A Saprykin[1], **Yu P Sharkeev**[2], **E A Ibragimov**[1], **E V Babakova**[1], **D V Dudikhin**[1]
[1]Yurga Institute of Technology, TPU Affiliate, Yurga, Leningradskaya str.26, Russian Federation
[2]Institute of Strength Physics and Materials Science of the Siberian Branch of the Russian Academy of Sciences, 2/4, pr. Akademicheskii, Tomsk, Russian Federation

E-mail: babakova@tpu.ru

Abstract: Alloys based on the titanium-niobium system are widely used in implant production. It is conditional, first of all, on the low modulus of elasticity and bio-inert properties of an alloy. These alloys are especially important for tooth replacement and orthopedic surgery. At present alloys based on the titanium-niobium system are produced mainly using conventional metallurgical methods. The further subtractive manufacturing an end product results in a lot of wastes, increasing, therefore, its cost. The alternative of these processes is additive manufacturing. Selective laser melting is a technology, which makes it possible to synthesize products of metal powders and their blends. The point of this technology is laser melting a layer of a powdered material; then a sintered layer is coated with the next layer of powder etc. Complex products and working prototypes are made on the base of this technology. The authors of this paper address to the issue of applying selective laser melting in order to synthesize a binary alloy of a composite powder based on the titanium-niobium system. A set of 10x10 mm samples is made in various process conditions. The samples are made by an experimental selective laser synthesis machine «VARISKAF-100MB». The machine provides adjustment of the following process variables: laser emission power, scanning rate and pitch, temperature of powder pre-heating, thickness of the layer to be sprinkled, and diameter of laser spot focusing. All samples are made in the preliminary vacuumized shielding atmosphere of argon. The porosity and thickness of the sintered layer related to the laser emission power are shown at various scanning rates. It is revealed that scanning rate and laser emission power are adjustable process variables, having the greatest effect on forming the sintered layer.

1. Introduction

Titanium and its alloys are widely-used materials for manufacturing medical implants and surgical instruments. Tooth replacement and orthopedic surgery are the fields, where these alloys are used most frequently (Figure 1) due to their good mechanical properties, bio-compatibility and high corrosion resistance.

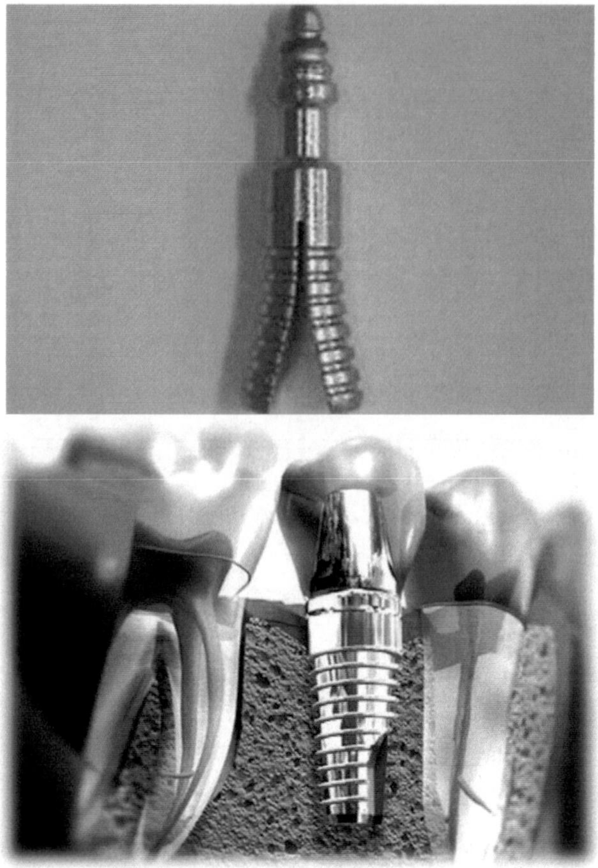

Figure 1. Tooth implant produced of titanium alloy [1].

The high modulus of elasticity (above 100 GPa) is a disadvantage of titanium and alloys on its base. Therefore, the main part of mechanical load is applied to the bones, and not to the implant, resulting in bone tissue resorption (Figure 2), implant weakening and reoperation [2]. At present, search and development of alloys with the low modulus of elasticity, having high strength and sufficient product plasticity though, are in focus of the scientists [3, 4].

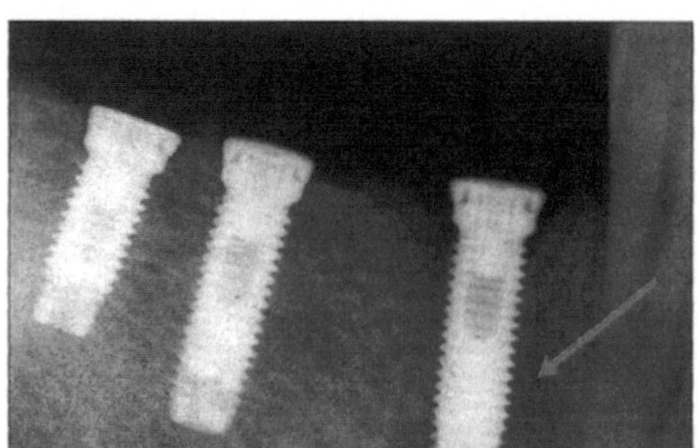

Figure 2. Resorption (failure) of a bone around the implant [5].

Alloys of the titanium-niobium system are distinguished by the most appropriate combination of mechanical properties. Moreover, niobium and titanium are non-toxic elements and do not react with body tissues. Some alloys of the titanium-niobium system have the low modulus of elasticity.

The conventional metallurgical methods of manufacturing titanium alloys are imposed by the technical and principal restrictions [6]. Titanium and cobalt-chromium alloy implants are manufactured via casting with further milling or turning, using methods of powder metallurgy, hot stamping, and stamping with further milling.

Casting is also a method to produce implants. High-frequency casting machines, providing reliable quality of casts, are used for melting steel or cobalt-chromium alloys. As a rule, producing implants of cobalt-chromium alloys is not difficult in laboratory conditions. However, casting implants of titanium-based alloys is more difficult because titanium reacts with nitrogen and oxygen when heating in air, and with form lining in the molten state. The technology of casting is based on vacuum melting and casting, and unconventional high fireproof forms are required as well. Therefore, titanium implants with a good quality surface can be produced in industrial conditions only. Cast implants are subject to X-ray diffraction control, revealing up to 5–8 % of wastes [7].

Laminar implants can be milled only when developing new titanium structures. In mass production milling an implant of a metal sheet is non-value-added. Spiral, cylindrical implants can be produced via turning only.

Hot stamping is a principal procedure of metal processing, including titanium and cobalt-chromium alloys. However, it should be noted implants are small products, and their gas absorbability grows when heating, therefore, technological procedures are strictly standardized. Equipment for stamping is quite complex and expensive, so it can be used when mass-manufacturing critical parts.

The procedure based on stamping and milling is technologically simple, so mass-manufacturing of products is more profitable. Test-production of titanium implants has revealed that it is the most appropriate and acceptable variant. The production of implants comprises two phases. The first phase includes cutting implant half-products by a special stamp from a 3 mm thick titanium sheet. The second phase is aimed at simultaneous thinning the implant base on both sides up to 1.2 mm by a milling machine. As burrs are removed, the implant is prepared for electrochemical polishing [7, 8].

These days the methods of layer-by-layer laser synthesizing end products of various powder materials are being successfully adopted by the health care system - one of the most important fields of human activity. It is mainly because the manufacturing cycle of the implant is shorter than that of conventional processing methods (milling, turning etc.), it is possible to save expensive materials and manufacture products of a complex geometry and specified porosity as well [9, 10].

Selective laser melting is a procedure of manufacturing a 3D product via successive laser beam fusion of powder materials according to a specified program (Figure 3).

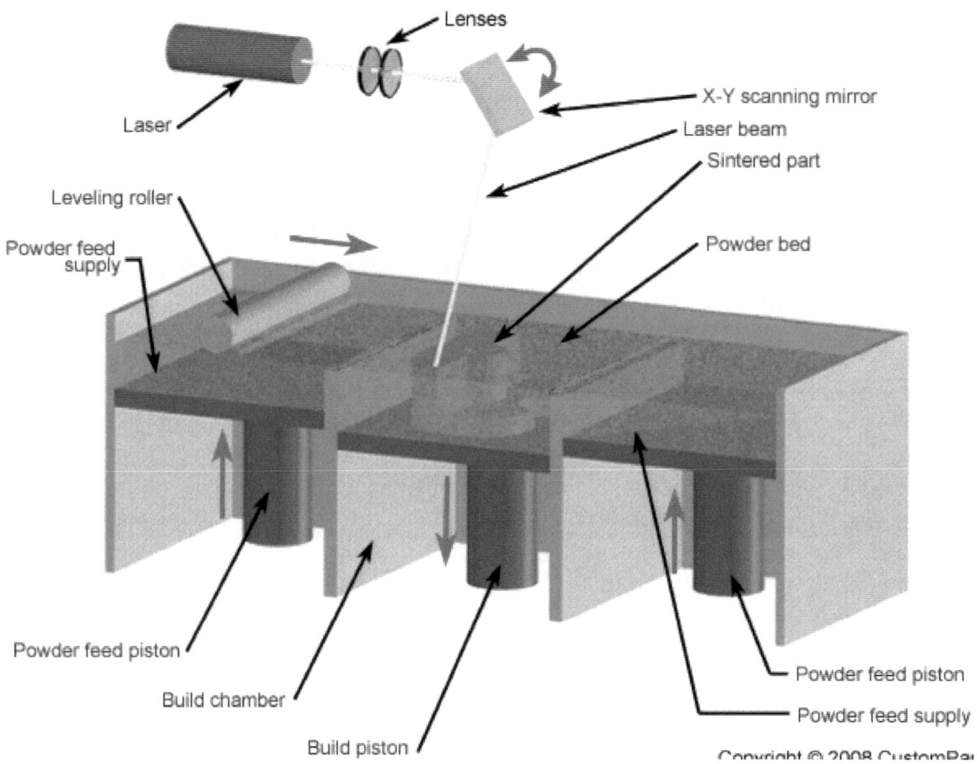

Figure 3. A layout of selective laser melting process [11].

Selective laser melting supports selecting and monitoring all basic parameters, which affect the process of product synthesizing such as emission power, scanning rate, and temperature of powder heating [11-13]. So it is possible to manufacture end products with specified mechanical characteristics [14].

The process includes two phases: first, an even thin layer of powder is spread all over the working area, then the laser is switched on; and the areas similar to the section of object to be produced are melted. A piston goes down approximately as deep as thick the layer is, then, the procedure is repeated, as soon as the top point of the model is approached.

Process parameters are adjustable at each phase of selective laser melting. To put the powder one can use a roller or a scraper. Melting is possible on the outer contour of the layer only or over the complete depth of the model.

Selective laser melting requires no supporting structure as a big amount of powder around the model keeps it from destruction, unless the end form is acquired and the mechanical article strength is achieved [15-17].

2. Methods and equipment of experimental research

A SLS machine «VARISKAF-100MB» is used for selective laser melting a composite powder containing 60 wt. % Ti and 40 wt. % Nb (Ti40Nb) [16]. This machine supports SLS/SLM synthesizing products of various physical configurations from a powdered material. The products can be sintered both in vacuum and in diverse gaseous mediums. The layout of the machine VARISKAF-100MB is given in Figure 4.

As the result of carried out experiments a set of samples is produced in the following process conditions. Laser emission power is 68, 75, 86, 96 and 106 W; scanning rate is varied in the range 0.017–0.05 m/s; scanning pitch is 0.1 mm.

Multilayer 10x10 samples are produced when thickness of powder layer is 0.7 mm. Samples are melted in argon. Plates made of alloy BT1-0B are used as a substrate. The substrate with powder is heated up to 200 °C, kept for 15 minutes in vacuum at the certain temperature, and melted in argon afterwards. The process temperature is controlled by a thermocouple of grade XA, the junction of which is in a substrate. Argon is supplied to the chamber as long as melting is carried out.

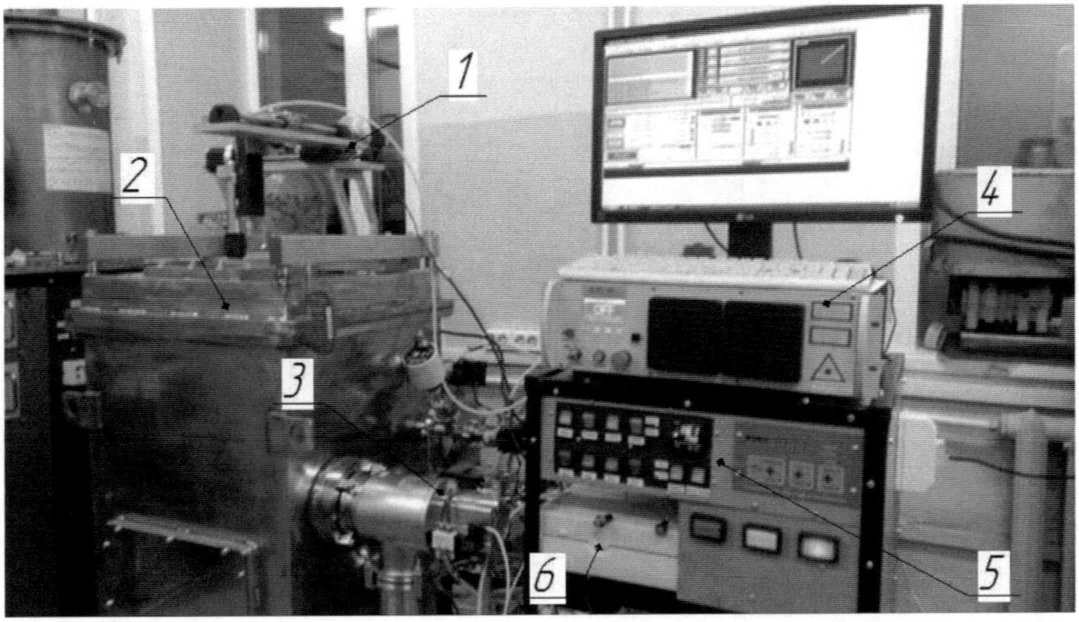

Figure 4. The layout of the machine VARISKAF-100MB: 1 – «flying» optical instruments; 2 – vacuum chamber; 3 – evacuation system; 4 – ytterbium laser; 5 – machine control assembly; 6 – computer numerical control.

An operating area, which shows preheating the substrate and melting the powder, is given in Figure 5.

Figure 5. Operating area of the machine.

Laser beam scanning strategy is a zigzag line (Figure 6).

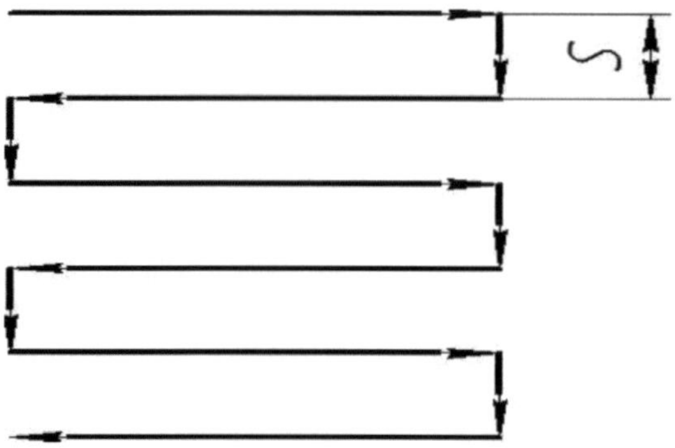

Figure 6. Scanning strategy.

3. Results and Discussion

Optical images of Ti40Nb samples are given in Table 1. As one can see, increasing scanning rate results in augmentation of quantity and dimensions of pores. Samples made at laser power 68–106 W and scanning rates 0.05 and 0.042 m/s do not differ much. However, the quantity of pores is decreased and uniform areas of melted powder are formed as the rate of scanning is reduced to 0.017–0.042 m/s and emission power grows up to 96–106 W (Figure 7).

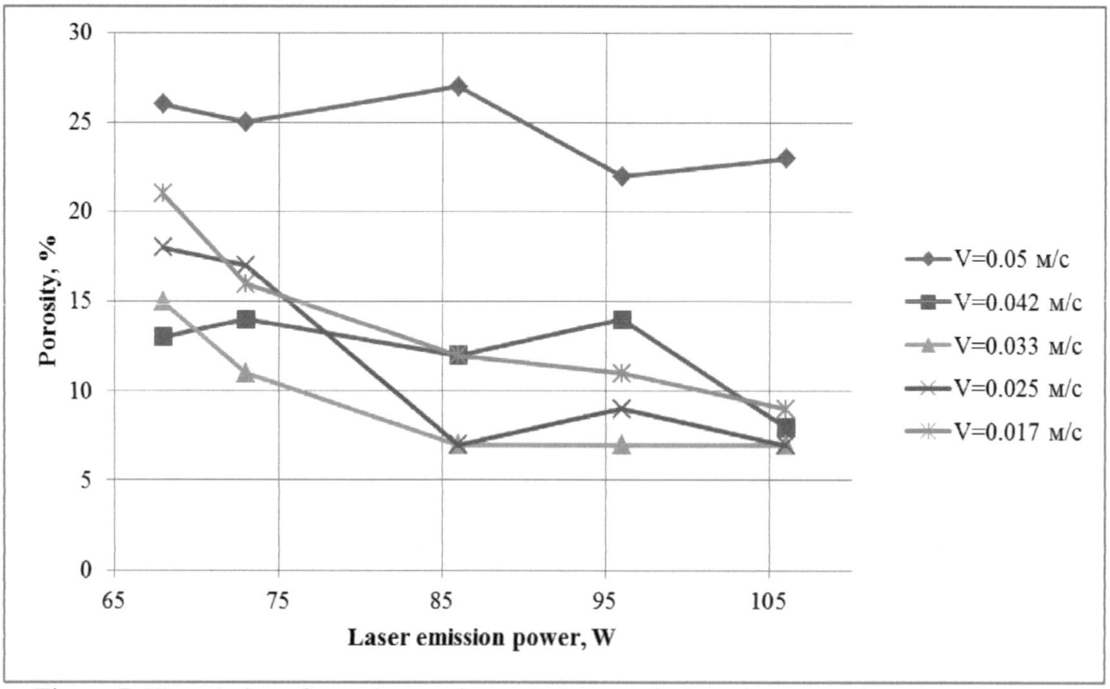

Figure 7. The relation of sample porosity to the laser emission power at various scanning rates.

The heat input grows and porosity decreases, as one can see in Figure 7, because of the increase in laser emission power and slowing down rate of scanning. Laser emission impact on the powder section takes longer as the rate of scanning is reduced, so fusion of the layer gets better.

Table 1. Single layers of alloy Ti40Nb

Rate, m/s	Laser emission power, W				
	68	75	85	96	106
0.05	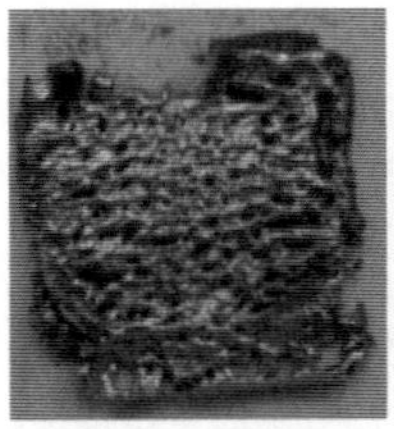	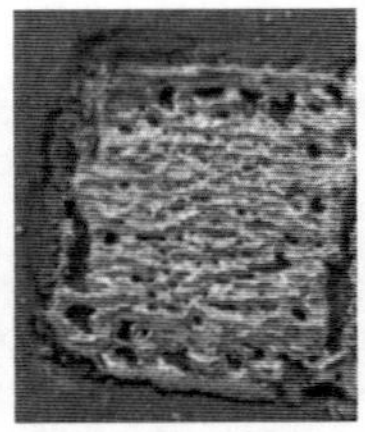	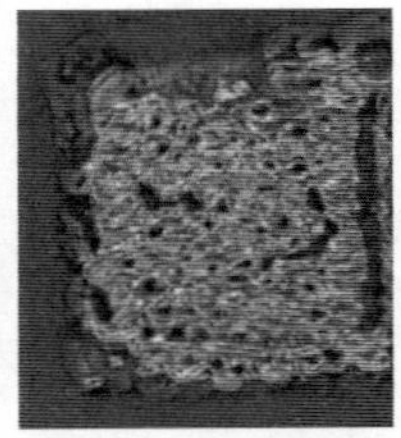	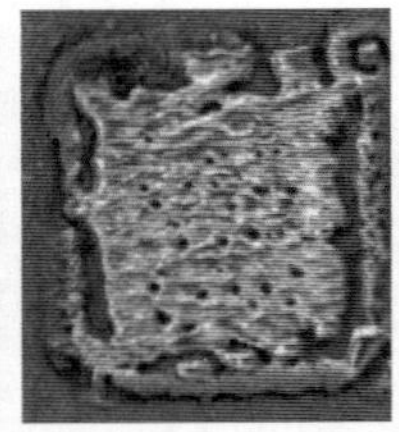	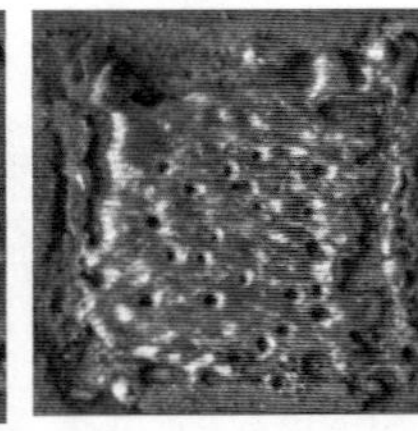
0.042	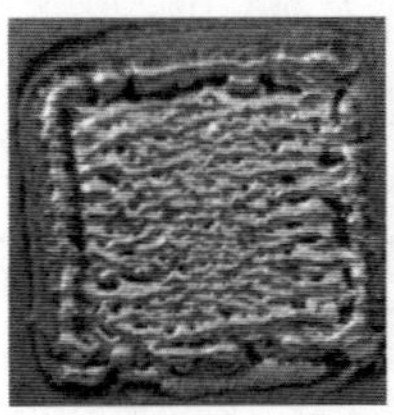	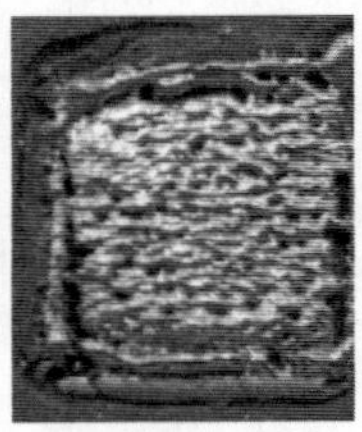	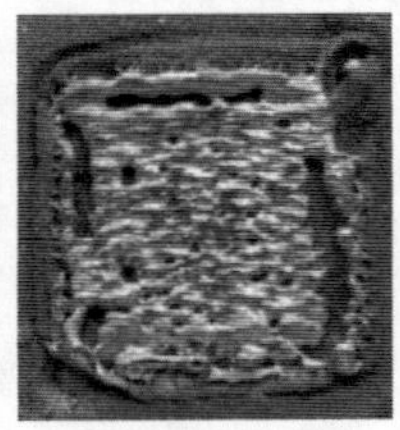	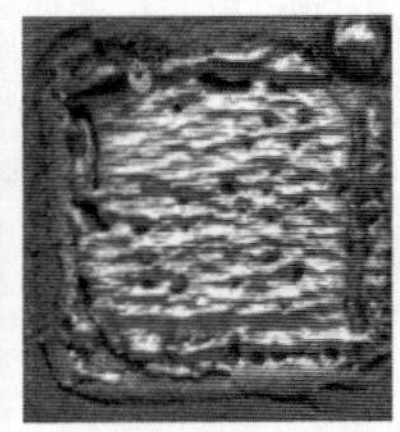	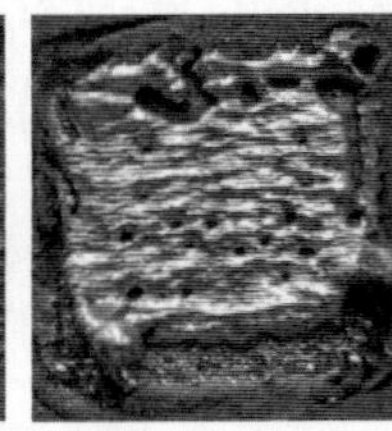
0.033	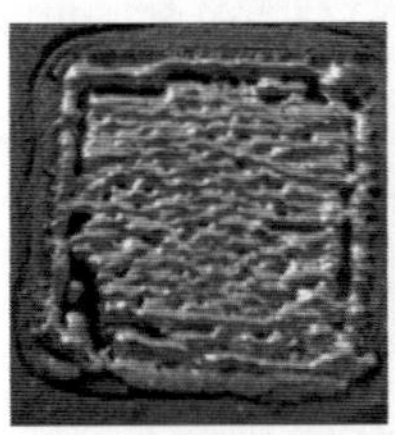	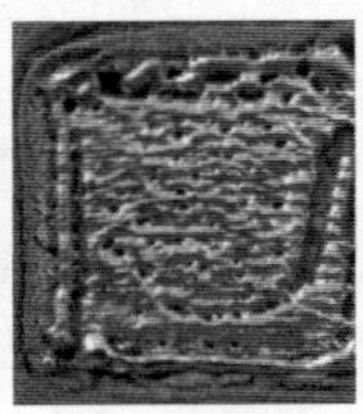	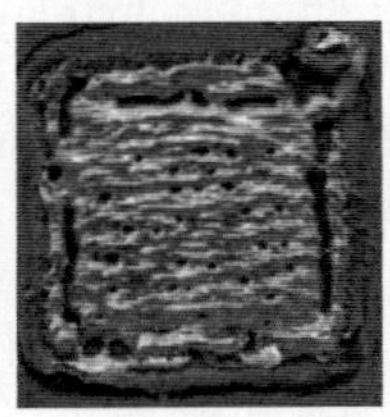	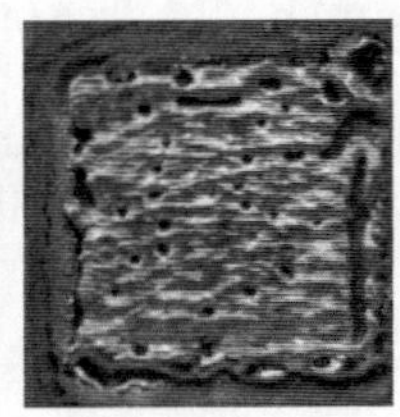	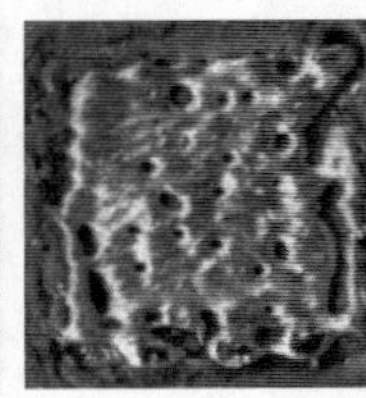
0.025	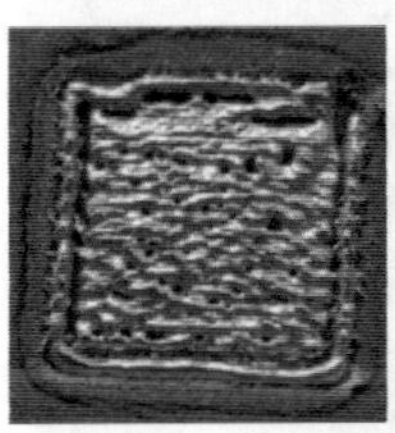	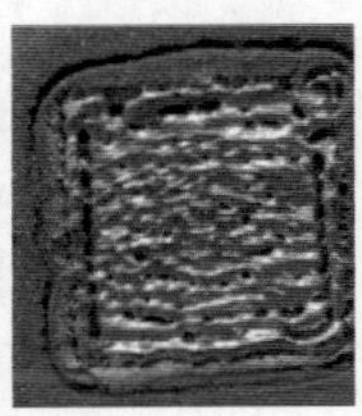	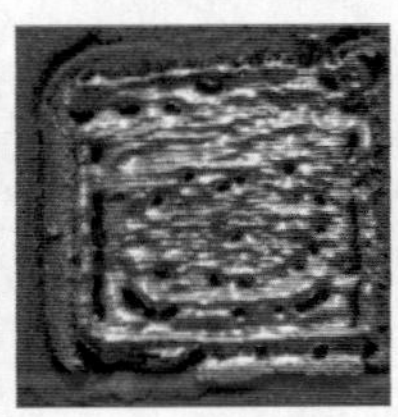	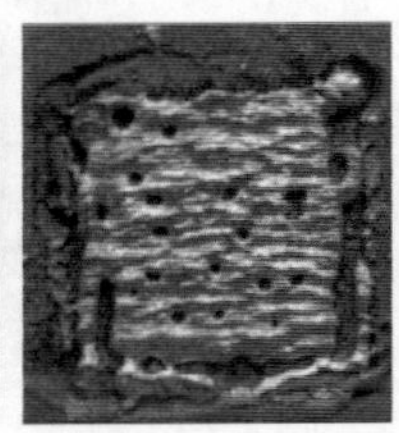	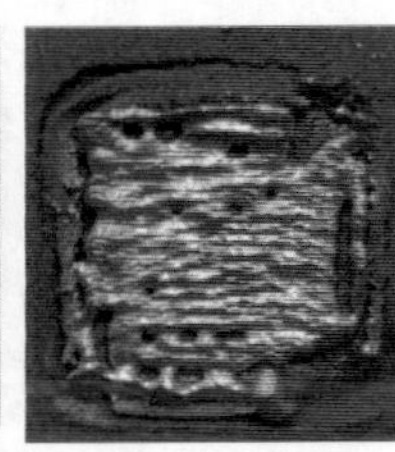
0.017	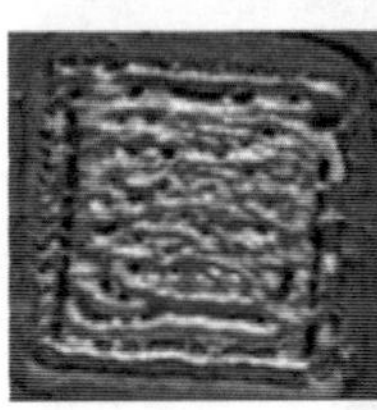	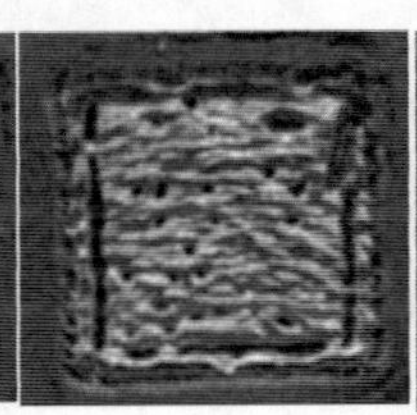	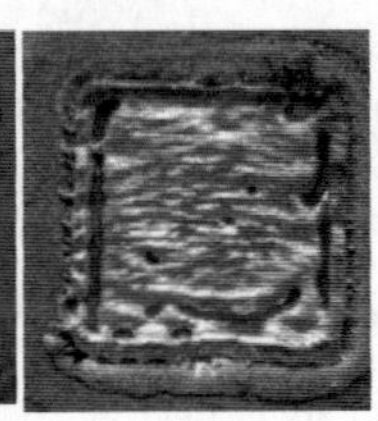	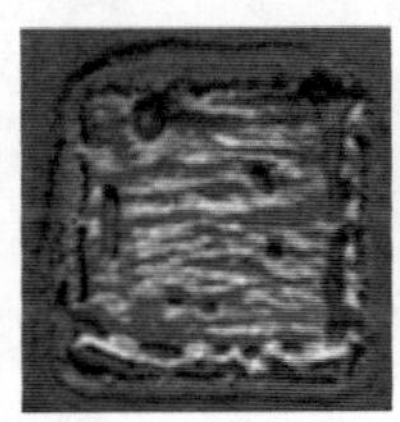	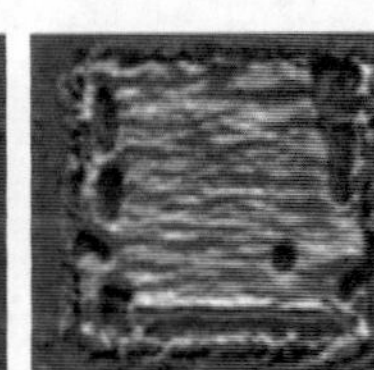

We measure the formed layer thickness in samples produced under emission power 106 W. In Figure 8 one can see that penetration depth is reduced as the rate of scanning is decelerated.

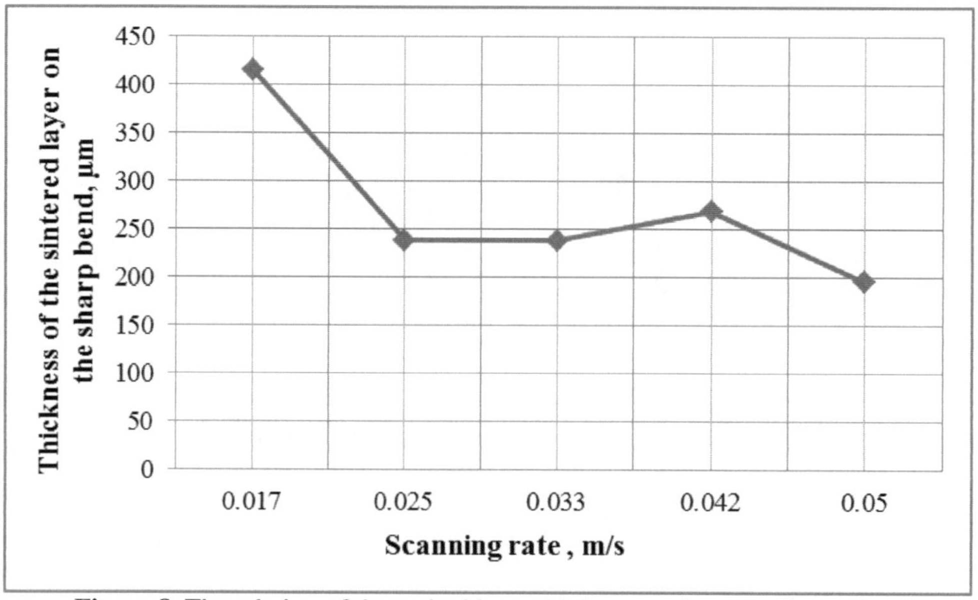

Figure 8. The relation of the melted layer thickness to the rate of scanning.

As the consequence, we carried out an experiment and made a sample in the conditions: laser mission power 106 WBт; scanning rate 0.017 m/s and pitch of scanning 0.1 mm. To increase the heat input a titanium substrate is pre-heated up to 400 °C. The sample produced in conditions of the maximally possible heat input of the machine is shown in Figure 9.

Figure 9. A melted layer of a composite powder in conditions of a maximal heat input.

The porosity of the sample is approximately 9 %, however, the thickness of the melted layer increases considerably to 570 µm, probably because of increasing temperature and the total heat input.

4. Conclusion

To sum up a conclusion can be drawn that laser emission power and rate of scanning are principal variables, which are important to form a melted layer of Ti40Nb composite powder. It should be noted that the slowing down rate of scanning and the increasing laser emission power result in the decrease of porosity and increase in the thickness of the formed layer of the alloy.

Acknowledgments
The work is performed under financial support of the Russian Scientific Foundation, grant *15-19-00191.*

References

[1] Sharkeev Y P, Eroshenko A Y, Kulyashova K S, Fortuna S V, Suvorov K A, Epple M, Prymak O, Sokolova V, Chernousova S 2012 Microstructure, mechanical and biological properties of zirconium alloyed with niobium after severe plastic deformation *J. Materialwissenschaft und Werkstofftechnik,* **44 (2-3)** 198–204
[2] Zhuravleva K, Scudino S, Khoshkhoo M S, Gebert A, Calin M, Schultz L, Eckert J 2013 *J. Advanced Engineering Materials* **4** 262–268
[3] Sharkeev Y P, Eroshenko A Y, Glukhov I A, Zhu Q, Tolmachev A I 2014 *J. Physical Mesomechanics of Multilevel Systems* **1623** 567–570
[4] Sharkeev Y P, Kovalevskaya Z G, Zhu Q, Khimich M A, Parilov E A 2015 *Advanced Materials Research* **1085** 307–311
[5] http://dentaltechnic.info/index.php/implanty
[6] Kuznetsov V E 2003 *J. CAD/CAM/CAE Observer* **4 (13)** 2–7
[7] http://dentaltechnic.info/index.php/implanty/zubnoe-protezyrovanye-na-ymplantatakh/2836-tyehnologiya_izgotovlyeniya_implanta
[8] Gibson I B, Rosen D W, Stucker B 2009 *Additive manufacturing technologies. Rapid prototyping to direct digital manufacturing* (New York, USA: Springer)
[9] Shishkovsky I V, Zakiev E S, Kholpanov L P 2005 SLS layer-by-layer synthesizing products of titanium nitride *J. The Physics and Chemistry of Materials Processing* **3** 71–78
[10] Saprykina N A, Saprykin A A 2012 Engineering support for improving quality of layer-by-layer laser sintering, *IFOST2012* **129–132**
[11] http://www.custompartnet.com/wu/selective-laser-sintering
[12] Kumar S, Kruth J-P 2009 Composites by Rapid Prototyping Technology *J. Materials & Design* **1–23**
[13] Ibragimov E A, Saprykin A A, Babakova E V 2014 Influence of Laser Beam Machining Strategy at SLS Synthesis, *Advanced Materials Research* **1040** 764–767
[14] Murali K, Chatterjee A N, Saha P, Palai R, Kumar S, Roy S K, Mishra P K, Choudhury A R 2003 Direct selective laser sintering of iron-graphite powder mixture *J. Journal of Materials Processing Technology* **136** 179–185
[15] Saprykin A A, Gradoboyev A V, Yakovlev V I, Ibragimov E A, Babakova E V 2015 Comparative analysis of processes to activate copper powder ПМС-1 for SLS synthesizing products *J. Metal processing* **3 (68)** 82–88
[16] Saprykin A A, Ibragimov E A, Babakova E V and Yakovlev V I 2015 Influence of mechanical activation of copper powder on physicomechanical changes in selective laser sintering products, *AIP Conf. Proc.* **1683** 020199
[17] Saprykin A A, Ibragimov E A, Babakova E V 2015 Engineering and Technology Based on Rapid Prototyping *J. Applied Mechanics and Materials* **770** 622–627
[18] Fischera M, Joguetb D, Robina G, Peltierc L, Laheurte P 2016 In situ elaboration of a binary Ti–26Nb alloy by selective laser melting of elemental titanium and niobium mixed powders. *J. Materials Science and Engineering* **62** 852–859

The synthesis of nanocomposite structures on the surface of geometrically complex products

B Brzhozovsky, V Martynov, E Zinina, M Brovkova, P Bochkarev
School of Electrical and Mechanical Engineering Yuri Gagarin State Technical University of Saratov, Saratov, Russia

E-mail: bmbsar85@mail.ru

Abstract. The paper presents the technology of nanocomposite structure synthesis in the surface layer of geometrically complex products under the impact of low-temperature plasma and the results of experimental study of their properties. An example is given of practical application of the nanocomposite structure product to improve the efficiency of ultrasonic dimensional processing.

1. Introduction

The current stage of production development in various fields is characterized by tougher working conditions for products made of metals and alloys. This requires continuous improvement of existing and development of new manufacture technologies in order to find the conditions allowing to take full advantage of opportunities for working at high temperatures, external power loads and corrosive environments.

One promising technology is the technology of forming composite structures on the working surfaces of products that ensure high and uniform hardness, wear resistance and dynamic strength [1-3].

Due to the low energy efficiency, traditional methods of gas thermal hardening and ion-plasma sputtering are time-consuming, which results in substantial thermal deformation of the whole product [4-6]. Creation of durable coatings based on conventional additive technologies with time dividing powder (or other material) application and subsequent intensive energy impacts do not ensure high dynamic strength of composite structures, including the bond with the matrix. In this respect, it is expedient to use the technology of composite structure formation directly within the surface layers of the products, i.e. in the original matrix. One such technology is the nanostructuring technology geometrically complex surfaces, which is carried out by high-performance multiplier effect on the interaction of electrostatic and electromagnetic fields [6]. Optimal interaction of the fields and environment properties is ensured by the special design of the plasma torch, the original waveguide system for the transmission of electromagnetic microwave energy from the magnetron to the product, the pumping system (vacuum of 300 Pa) and the supply of the process gas mixture. As a result, in the plasma torch low-temperature plasma is formed that provides intensive heating only of the workpiece surface layer, which leads to the formation of a more complex than the original structure. It is comprised of the amorphous phase based on particles $<$ 10 nm and nanoclusters of 30-70 nm. This provides high dissipative properties, strength and surface hardness. The transition layer between the amorphous layer and the matrix ensures the strength of the entire structure.

2. The results of experimental studies of the synthesis process

IOP Conf. Series: Materials Science and Engineering **140** (2016) 012002 doi:10.1088/1757-899X/140/1/012002

The study of the formed nanocomposite structure included examination of its morphology, chemical composition as well as its physical and mechanical properties. The research was carried out on P6M5 steel and RX10 carbide samples treated at the minimum and maximum parameter value combinations of electromagnetic and electrostatic fields with the use of analytical complex on the basis of a scanning electron microscope TESCAN MIRA\\LMU, double microscope MIS-11, analytical balance and mechanical property precision tester NANOVEA Ergonomic Workstation.

Examination of the morphology of the steel workpiece allowed to identify a ~2 μm thick layer on its surface which clearly shows the areas around the carbide grains containing numerous flat irregularly shaped particles ~100 nm thick and arranged in the form of a sphere or cloud (Figure 1, a). Outside the layer the number of particles is reduced considerably. There were also detected 350-550 nm inclusions of modified shape (from round to elliptical), the occurrence of which is possible only in the event of dissolution of the carbide grains and / or splitting them into smaller ones (Figure 1, b). The maximum number of inclusions (65% of total) are detected at the depth approximately equal to the average radius of undissolved grains, which corresponds to 0.7-0.8 μm, and the minimum (35% of total) may be located at their border, i.e. at the depth of 1.4-1.6 μm. The presence of the meniscus (a layer in the molten state) around the carbide grains in the region of their interface with the base material is the most characteristic feature of underlayer formation.

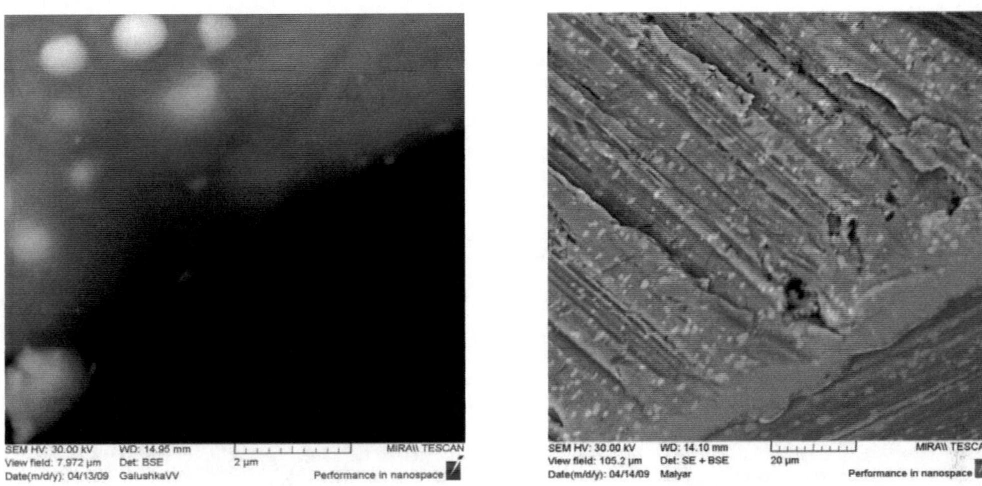

a b

Figure 1. A cloud of nanoscale particles (a) formed around partially dissolved carbide grain in the surface layer of the sample workpiece after treatment in a low-temperature plasma of combined discharge (b).

Examination of the surface morphology of the steel workpieces revealed the formation of nanoclusters on them. Clusters and larger structural elements of the surface (matrix grains) are joined into a whole by amorphous binder (Figure 2, a). Surface morphology analysis has also found that the plasma impact on them is carried out by groups of Ø40-50 nm nanoscale beams. In defect-free surfaces the traces of these groups are homogeneous and seen as solid virtually indistinguishable homogeneous zones. On the surfaces with different relief defects, such as irregularities of droplet phase type (Figure 2, b), microprotrusions, etc, the formation of beam focusing lenses takes place. The consequence is the electrostatic field distortion which leads to plasma uniformity breach and manifests itself in layering and regrouping beams towards their increased concentration to the extent of filamentation.

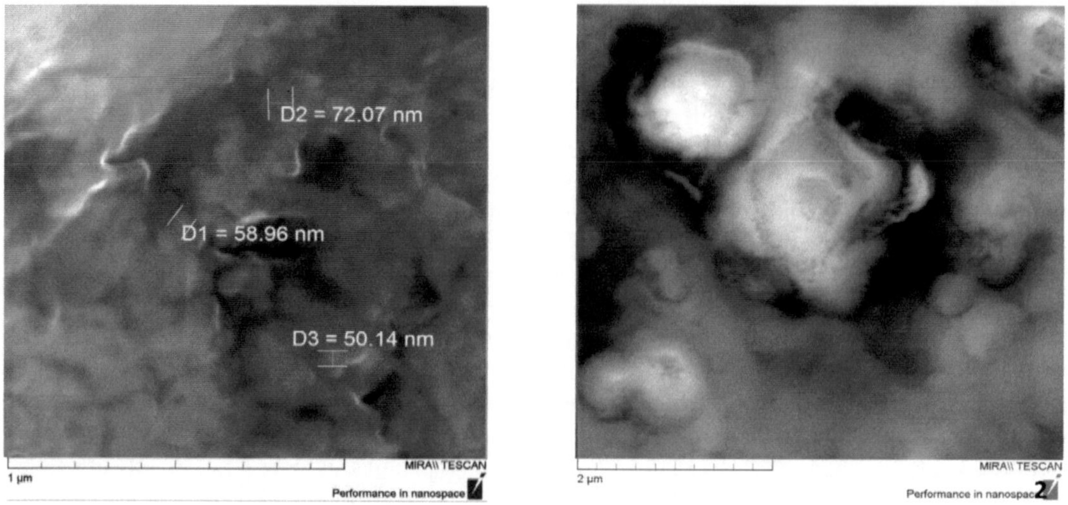

a b

Figure 2. Electron microscopic image of the carbide sample surface:
a – nanoparticle inclusions; b – droplet phase.

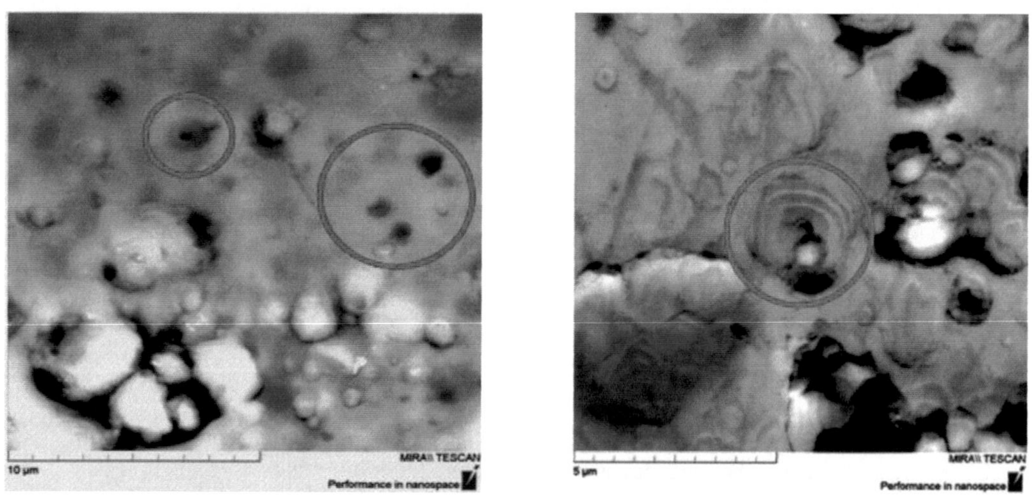

a b

Figure 3. Electron microscopic image of the carbide sample surface in the zones of blind holes, reflow (a) and sagging (b).

The impact of such beams leads to the formation of micron reflow zones and blind holes of 0.7-1.0 µm diameter (Figure 3a). The intense local heating in reflow zones in combination with plasma layering causes uneven volumetric heat release and thus, uneven heating of the surrounding areas, which leads to surface subsidence and the formation of nano-sized steps or closed terraces which outwardly appear as a nested set of closed curves (Figure 3, b) as well as to the formation of small clusters, reducing the amount of the amorphous phase and, consequently, a marked embrittlement of the surface.

Chemical composition study was carried out on the hard alloy samples using X-ray sensor INCAPentaFETx3, OXFORD Instruments, which is part of an analytical complex TESCAN MIRA\\LMU. The depth of the analyzed layer was about 1 μm. The initial structure of the sample surface layer consisted of aluminum (13.83 wt.%), titanium (38.01 wt.%), carbon (16.87 wt.%), nitrogen (27.96 wt.%) and oxygen (3.35 wt.%).

The comparison of the chemical composition of the sample surface fragments treated in the minimum plasma exposure modes shows that its local changes are caused by the processes occurring at the surface, and are associated with the changes in the structure of the plasma flow emitted by the inner edge. Thus, the flow impact on the tops of large droplets (Figure 4, a) is different from the peripheral sections (point 5), where reduction of the nitrogen concentration and its alignment with carbon and aluminum concentrations are registered. Furthermore, slight fluorine concentration was found, which means sublimation of fluoroplastic insulation of the sample support.

Similar changes were observed in the vicinity of the melt zone (Figure 4, b). In the area of maximum impact of the beam (point 4) there is maximum (54% from baseline) reduction in the nitrogen concentration and an increase in the concentration of carbon and titanium, which indicates substitution processes and/or carbon-nitrogen displacement. This eventually leads to nitrogen, carbon and aluminum concentration leveling off.

Exposure to plasma, leading to subsidence surface processes, is accompanied by changes in terms of convergence of titanium and nitrogen concentrations, as well as aluminum and carbon (Figure 4, c).

Thus, the change in the chemical composition of the surface layer in the minimum modes of plasma exposure is mainly due to the polymerization processes of island films of $Al_{15}C_{17}N_{21}O_5$-type formation, and layering structure processes can only be observed in the area of surface subsidence.

Changing modes to maximum plasma exposure has caused changes in the chemical composition (Figure 4, d, e) both similar to the effect on the minimum modes and new, which are characterized by the complete replacement of the nitrogen with nanostructures of $Ti_{57}C_{34}Al_7O_2$-type formation (Figure 4, f).

The study of physical and mechanical properties of the structure included a quantitative and semi-quantitative measurement and evaluation of its roughness as well as hardness investigation.

When measuring the roughness of steel specimens, initially the values of each trough and ridge of microroughnesses located in the linear field of the microscope were measured, and then the mean and standard deviation (Ra) from it was calculated. The obtained values show (Table 1) that, firstly, the roughness is reduced (2.93 times on average), and secondly, the degree of reduction depends on the initial state of the surface, i.e. on its technological heredity.

Measuring roughness of hard alloy samples was carried out on a lapped hardened plate surface (12 roughness class). After the installation of the specimen on the plate, a loose strap of thin silk thread with a light cup of thin plastic (0.3 g) at the end was put on the specimen. The starting force was determined with the help of some drops of water in the cup. When the load reached the value that balanced friction force, the specimen was displaced and began its movement. The amount of this load was determined by weighing to the nearest 0.01 g.

For each sample the average starting force was determined by 10 passages, both before and after plasma exposure. The results showed (Figure 5) that after low temperature plasma exposure the breakaway force decreases on average 1.4 times, due to microroughness blunting and reducing the height and number of contact points of engagement between vertices.

The study of electron microscopic images allowed to specify the mechanism and to identify the causes that lead to the reduction of the height of the microscopic irregularities. They are associated with the protruding microdefects deformation and their transformation into planar defects. The processes of film formation and changes in the chemical composition of the surface layer which change the conditions of contact interactions at the atomic level play an important role here.

The study of hardness was performed with loading the indenter (Berkovich pyramid) with weights of 50 and 100 mN. Processing of the obtained data (Figure 6) included the construction of the distribution of hardness and elastic module values as a result of the decomposition of print depths range into intervals, followed by counting the number of parameter values have fallen in each interval, their averaging and dividing each array members by its maximum value.

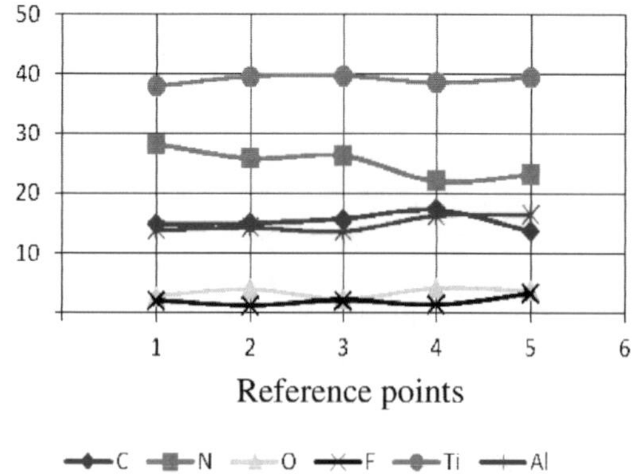

a

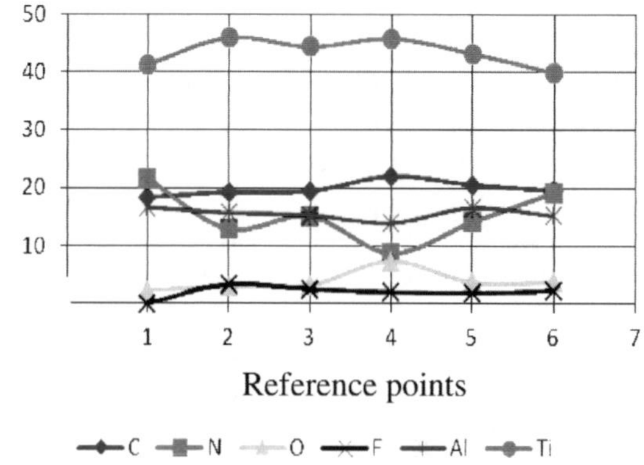

b

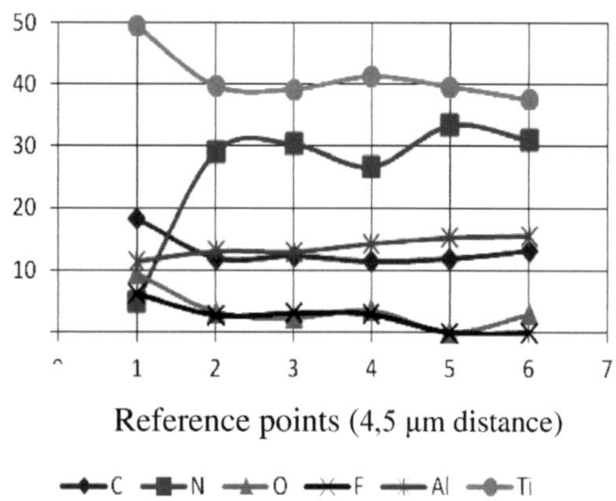

c

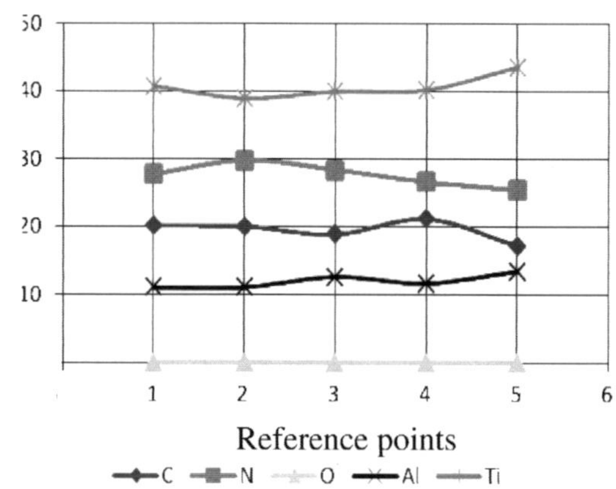

d

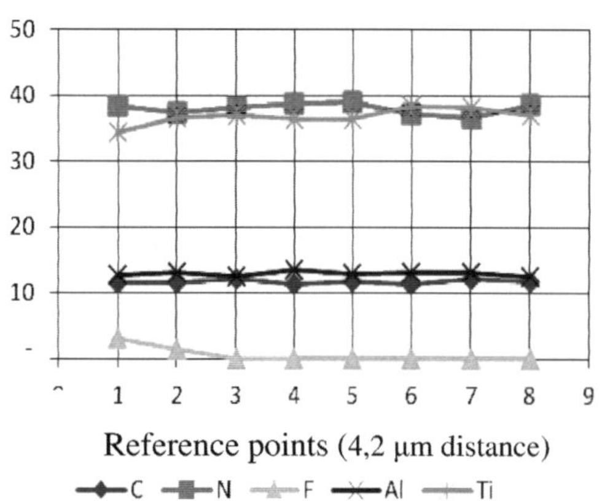

e

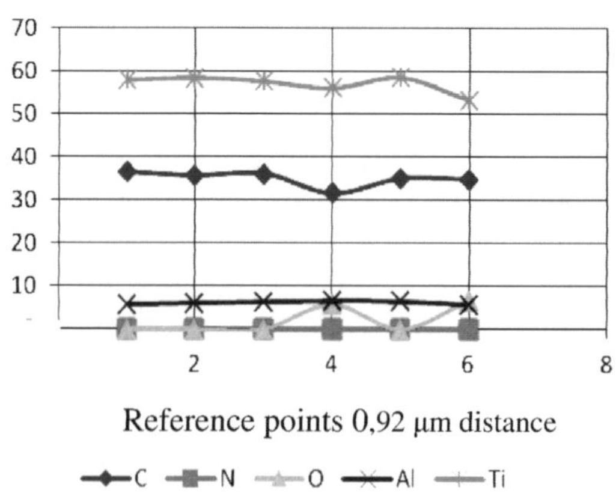

f

Figure 4. The chemical composition of the surface of the cemented carbide samples when exposed to plasma at the minimum (a-c) and the maximum (d-f) modes.

Table 1. The roughness of the surface of steel sample workpieces before and after combined discharge low-temperature plasma exposure

No	Roughness by Ra, μm		Reduction Ra, times
	initial	after treatment	
1	0.68	0.20	3.4
2	0.41	0.14	2.9
3	0.50	0.22	2.3
4	0.33	0.19	1.7
5	0.21	0.18	1.1
6	0.40	0.06	6.6
7	0.47	0.09	5.1
8	0.24	0.12	1.9
9	0.20	0.19	0.9
10	0.48	0.16	3.0
11	0.40	0.12	3.3
10	1.10	0.46	2.3
13	0,55	0.15	3.7
14	0.34	0.11	3.4
15	0.36	0.15	2.4

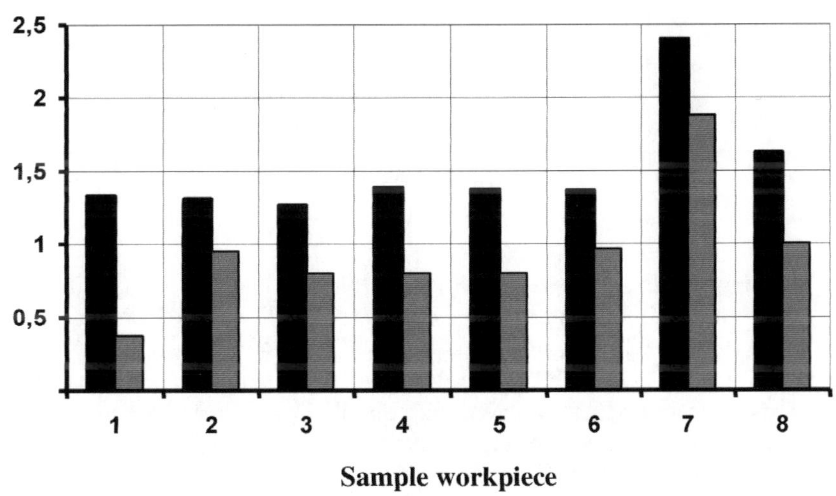

Figure 5. Comparative charts of breakaway effort of hard alloy samples before (black) and after (gray) low-temperature plasma exposure.

The analysis of the distributions confirmed that formed upon low temperature plasma exposure structure is indeed nanocomposite as its surface layer has zones with increased hardness as well as with high elasticity (Figure 7).

3. Practical implementation

As an example of the technology practical use let us consider the process of ultrasonic dimensional treatment of products made of quartz glass, based on the interaction of three main components: a working tool (contactor), powder (working slurry) and the workpiece.

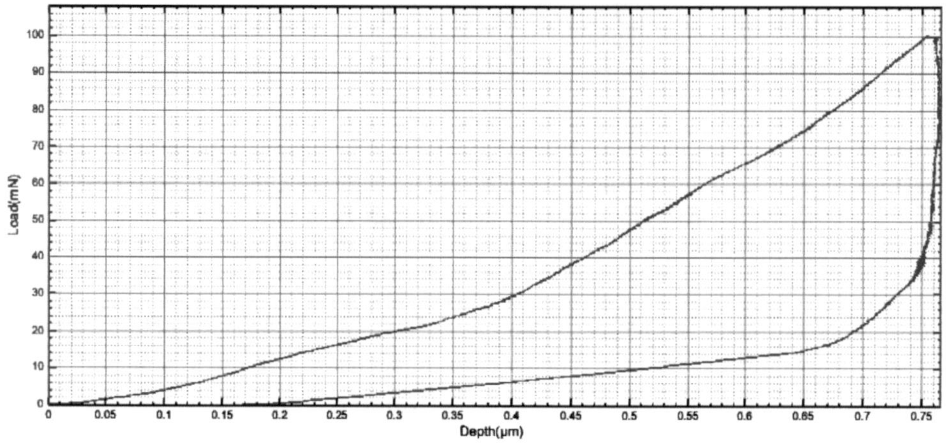

a

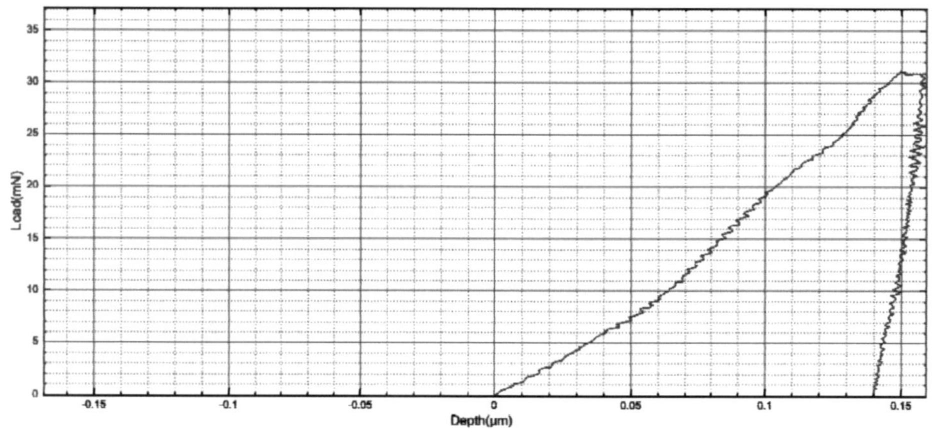

b

Figure 6. Examples of 'loading – unloading' curves corresponding to the maximum values of the elastic (a) and hardness (b) modules.

Contactor vibrations are transmitted to the working suspension that performs the functions of the cutting tool. Boron carbide powder, pre-mixed with distilled water in a certain ratio was used as slurry. During the processing, boron carbide grains penetrate the processed material with their sharp edges and facets, while the contactor acts as a 'hammer', regularly striking the powder.

It should be noted, however, that along with sufficiently high efficiency of this processing method, there are several problems associated with rapid mechanical wear of the contactor working parts (Figure 8). The main consequence is the deterioration in the geometry (tapering) and the roughness of working surfaces of manufactured products. In this regard, it would be relevant to increase strength and wear resistance of the surface layer of the contactor working part through its nanostructuring, which can further increase the efficiency of ultrasonic treatment process in term of ensuring the specified accuracy and cost reduction.

The quality of nanostructuring process was evaluated through quantitative assessment of changes in physical and mechanical properties of the surface layer of the contactor working part material according to microhardness measurement (Vickers) both before and after low-temperature plasma exposure using a technique based on determining of microhardness distribution in the depth of the surface layer [7].

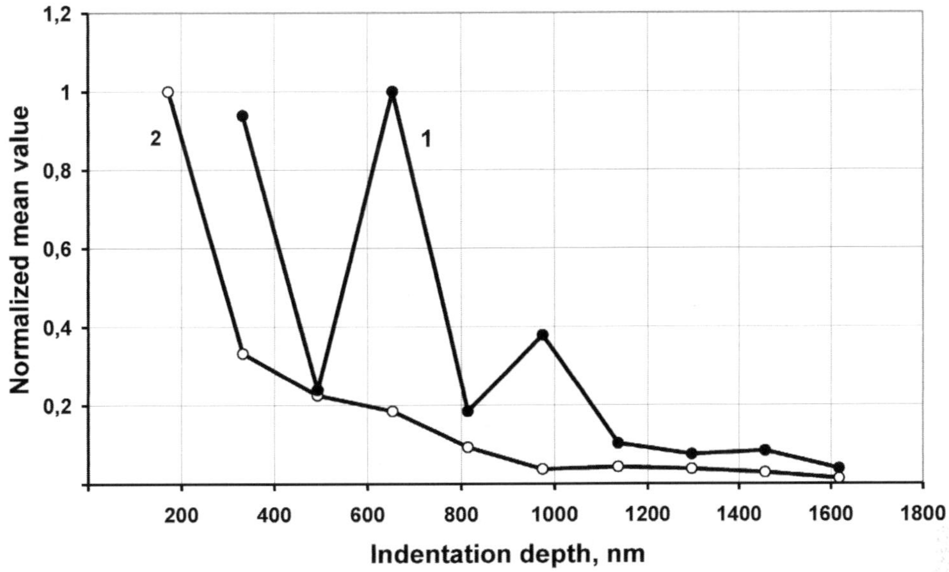

Figure 7. The distribution of the elastic module (1), and microhardness (2) in the modified layer.

Figure 8. Contactor.

The measurements of microhardness of three contactors and their comparative analysis confirmed the following:

1. The least solid is the part of the surface layer that is located directly below the adsorbed zone and the zone of the oxides. This is due to the fact that the contactors were subjected to abrasive treatment to form the desired geometry of the working part after manufacturing. As a result, elastoplastic deformation and heating caused extension of the surface layer, irreversible structural and phase changes therein, and subsequent decrease of microhardness.

2. Contactor №1, previously used, was subjected to repeated abrasion with a view to restoring the geometry of the working surfaces. After plasma exposure, their microhardness increased by 1.35 times on average. Operating results made it possible to record the growth of the contactor resistance time by 2.5 times compared to the initial data.

3. Microhardness of the new contactor №2 after plasma exposure increased by 1.25 times on average; resistance time increased by 2 times. It was found that degree of dispersion and boron carbide powder grain shape influence the resistance. The effect is that during processing of the first two workpieces, interaction with the contactor decreases the grain size of the initial fraction from about 35 μm at their distribution center to about 30 μm, and their shape is changed from irregular polygonal to spherical one. It is during this period the contactor wears the most, creating, however, favorable conditions for further processing. This means that there is an additional reserve for increasing time resistance, which can be used in the following way. The transformation of boron carbide original structure occurs in the processing of the first workpiece with a conventional contactor. As a result, a defect layer which exceeds the initial microhardness is formed on its surface. This is a favorable factor from the viewpoint of low-temperature plasma exposure. Thereafter the contactor undergoes plasma exposure, the duration of which does not exceed 15 min., and then it is placed into position for the main processing cycle (note that the installation time is small and achieved through the technologically effective contactor construction).

4. The use of the proposed approach to contactor №3 showed that after plasma exposure its working part surface layer microhardness increases by 1.35 times on average, which is 1.46 times higher than the microhardness of contactor №1. As a result, the resistance time of contactor №3 increased by 3 times.

4. Conclusion

The presented data show that the quality of nanocomposite structure synthesis on the surface of the geometrically complex products is closely associated with the method and the specifics of the prior exposure of the workpiece material changing the properties of their surface layers. All previous machining operations (turning, grinding, polishing with diamond paste) and heat treatment (hardening and annealing) will have different, in terms of magnitude and speed, elastic-power and thermal effects on the surface layer, as each subsequent operation is performed on the already modified surface. In macroaspect, it leads to fluctuations in the surface roughness and the formation of unstable strengthening parameters called wear hardening, which cause significant changes in the structure and properties of the surface layer. Moreover, these changes are random and can be of either positive or negative character. However, the energy of charged low-temperature plasma particles, affecting the surface layer changes its structure from the original high defect structure into a composite one having a higher affinity to the base material, and on this basis provides increased main indicators of operational reliability of products, in particular, indicators of failsafety.

Acknowledgement

The authors would like to acknowledge the assistance of the Russian Science Foundation. The present study was supported by the Russian Science Foundation grant (Project №15-19-00030).

References

[1] Fedotov A, Agabekov Yu, Machikin V 2008 Multifunction nanocomposite coatings, *Nanoindustry*, **1** 24-26 (in Russian).

[2] Musil J 2000 Hard and superhard nanocomposite coatings, *J. Vac. Sci. Technol.* **125** 322-330.

[3] Veprek S 1999 The search for novel, superhard materials, *J. Vac. Sci. Technol. A.* **17(5)** 2401-2420.

[4] Bakhvalov Yu O, Mishenznikov G E, Anisimova A S et al. 2011 Synthesis and study of production of antifrictional high-resistance coating for mechanical engineering using magnetron deposition of chromium and ion implantation, *Composites and nanostructures* **4** 5-13 (in Russian).
[5] Musil J, Vlcek J 2002 Physical and Mechanical Properties of Hard Nanocomposite Films Prepared by Magnetron Sputtering, *Proc. 5 Int. Conf. on Modification of Materials with Particle Beams and Plasma Flows* (Tomsk: Scientific conference proceedings) **3** 393-398.
[6] Pashentsev V 2010 Vacuum ion-plasma deposition of nanostructured coatings, *Nanoindustry*, **5** 26-28 (in Russian).
[7] Sergeev A A, Zinina E P, Kislitsyna N F 2006 The method and apparatus for receiving the stationary combined discharge of low-temperature decreased pressure plasma Patent №2277763 Russia (in Russian).
[8] Brzhozovskii B M, Zinina E P, Martynov V V, Pleshakova E S 2015 Assessing the quality of the surface layer of the cutting tool working part in microhardness parameter, *Metalloobrabotka*, **2 (86)** 15-21 (in Russian).

Multiprocessing and Correction Algorithm of 3D-models for Additive Manufacturing

R R Anamova, S V Zelenov, M U Kuprikov, A V Ripetskiy
Moscow Aviation Institute (National Research University),
Moscow, Russia

E-mail: anamova.rushana@yandex.ru

Abstract. This article addresses matters related to additive manufacturing preparation. A layer-by-layer model presentation was developed on the basis of a routing method. Methods for correction of errors in the layer-by-layer model presentation were developed. A multiprocessing algorithm for forming an additive manufacturing batch file was realized.

1. Introduction

The emergence of new 3D-printing technologies has resulted in a dramatic increase in hardware and software development [1-3]. Printing capabilities, material specifications and accuracy of modern industrial additive manufacturing equipment, all point at the feasibility of using such technologies for production of different engineering items. At the same time, additive manufacturing preparation software remains an open issue, due to the fact that modern additive manufacturing preparation software is not multipurpose [4].

Software systems for additive manufacture preparation have the following weaknesses:

- costs of material and computing resources;
- impossibility to select algorithms for error correction;
- impossibility of editing, control of sintering layers and checking of item topology after slicing;
- high probability of errors during files transfer from working chamber arrangement software to the machine control program generating software;
- impossibility of working with operating systems other than Windows and other hardware platforms.

In this connection, the development of new additive manufacturing preparation software remains a pressing problem.

Preconditions for the development of additive manufacturing preparation software include:

- formulation of requirements and identification of constraints for automation tools of layer-by-layer synthesis method for manufacture preparation;
- definition of mathematical and algorithmic support;
- verification of the developed mathematical support and software.

2. Development of layer-by-layer model presentation

During manufacture preparation process, discrete presentation of the model with 3D-printer precision is generated. We can describe the model surface as a set of triangles which are identified by three ordered vertices.

Empirically, the following model validity criteria were formulated:

- the surface should be continuous;

- the surface should confine the space;
- volumes confined with various surfaces should not overlap;
- different surfaces should not contain the same points.

Validity criteria may be checked using the routing method.

2.1. Routing method

Geometrical model is presented as a grid or a number of connected bounds (Figure 1). We can choose any point of space $\tilde{o}_0$ and vector $\overline{p}$ and consider intersection points of the resulting line and model bounds.

Let us consider intersection points for line $L(\overline{p}, x_0)$ and model bounds. Let us assume that $\overline{v_i}$, $i = [1...n]$ are bounds, which are intersected by line $L(\overline{p}, x_0)$ in points $\tilde{o}_i$. Since bound $\overline{v_i}$ is an ordered set of points, we can calculate a normal to the bound as a cross product:

$$\overline{v_i} = [x_1, ..., x_3], \quad \overline{p_i} = [\overline{p_{i1}}, \overline{p_{i2}}] = [x_2 - x_1, x_3 - x_1].$$

All bounds intersected with line $L(\overline{p}, x_0)$ can be divided into exterior and interior ones. Bound $\overline{v_i}$ will be exterior for vector $\overline{p}$, if the angle between $\overline{p_i}$ and $\overline{p}$ is less than 90°. We can define it as a cross product:

$$v_i = \begin{cases} exterior, if \left(\overline{p_i}, \overline{p}\right) = \left|\overline{p_i}\right|\left|\overline{p}\right|\cos(\phi) > 0, \\ interior, if \left(\overline{p_i}, \overline{p}\right) = \left|\overline{p_i}\right|\left|\overline{p}\right|\cos(\phi) < 0 \end{cases}$$

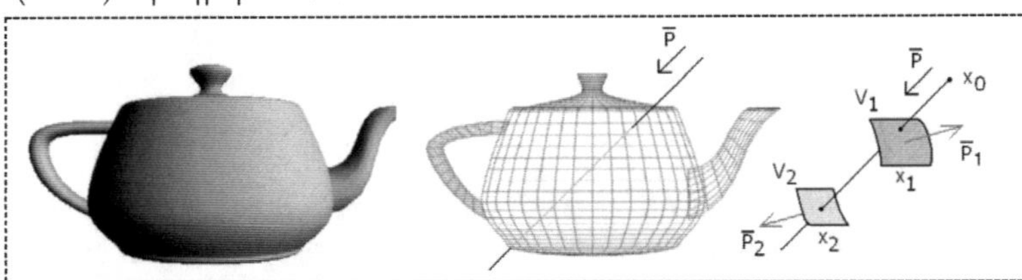

Figure 1. Model (left), grid presentation of the model (center) and routing (right)

Let us compose set $\grave{I}$ from pairs $(x_i, \overline{p_i})$, $\tilde{o}_i$ is intersection point of bound $\overline{v_i}$ with line L, $\overline{p_i}$ is normal vector of bound $\overline{v_i}$. Points $\tilde{o}_i$ are ordered on line L. Model is closed relatively to line L, if the first bound is exterior and orientation of each following bound is different from that of the previous bound. Using scalar product: a set of $\grave{I} = \left(x_i, \overline{p_i}\right)$ pairs, $i = 1, ..., n$ is valid, if $(\overline{p_0}, \overline{p}) > 0$, $(x_0, x_i) < (x_0, x_{i+1})$, $(\overline{p_i}, \overline{p}) = -(\overline{p_{i+1}}, \overline{p})$.

Validity of the model as a whole may be formulated as follows: model V is valid, if no $\overline{p}$ vector and $\tilde{o}_0$ point exist, for which the set of $\grave{I} = \left(x_i, \overline{p_i}\right)$ pairs, $i = 1, ..., n$ is not valid. We should check validity of the model only for the direction, from which its layer-by-layer presentation was calculated, and with resolution, which is necessary for manufacturing using particular equipment.

2.2. Generation of layer-by-layer model presentation

Layer-by-layer presentation of model V_i can be obtained using its subset in discrete space Z^3. Discrete model presentation (Figure 2) is based on function R: $R(V_i) \subset Z^3$. Using set of $M\left(x_{ij}, \overline{p}\right)$ pairs, which

result from intersection of the model with $\overline{p}$ line and contain point x_{ij}, we can form function R, which maps a set of M pairs in discrete space.
Let us define D_i as model V_i presentation in discrete space of whole numbers: $D_i = (d_1,...,d_n), d_i \in Z^3, D_i \subset Z^3$.

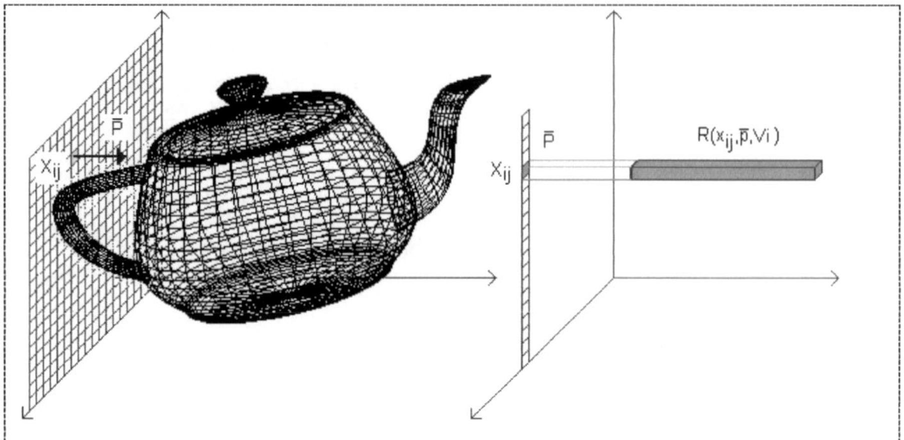

Figure 2. Parametrical and discrete model presentation

Discrete model presentation is used as input information for generation of numerical control commands; it can be obtained using the routing method for the set of x_{ij} pairs.

$$D(V_i) = \sum_{i,j} R(x_{i,j}, \overline{p}, V_i), D(V_i) \subset Z^3.$$

3. Development of methods for correction of errors in layer-by-layer model presentation

All additive manufacturing technological processes use layer-by-layer model presentation, which is realized through routing methods. When we use routing method algorithms, the input information includes discrete model presentation (triangles); the output information is layers. ‘Layer’ means vector or raster presentation of a model section in selected scanning plane.
Vector presentation can be obtained analytically by searching model sections in selected scanning planes. Raster presentation is a result of repetitive application of model triangles and scanning vector intersection points searching operation. Searching operation is repeated until the result with necessary precision is found.
The result of model routing is an ordered set of coordinates of the intersection point and normals in each point:

$$P(V_i) = [x_i, N_i], i = 1...n, x_i > x_{i-1}.$$

Using scanning vector direction and face normal, we can define the face type (outer or back) by scalar product sign. If we know outer and back faces of the model, we can identify errors in the geometry of a 3D-model.
In routing method, geometrical model contains no errors, if the ordered set of intersection points and appropriate normals contains first outer face, last back face, and faces interchange.
For example, the following set of points and normals does not contain routing errors (Figure 3) ($x_i < x_{i+1}$):

$$\{x_1, outer\}, \{x_2, back\}, \{x_3, outer\}, \{x_4, back\}.$$

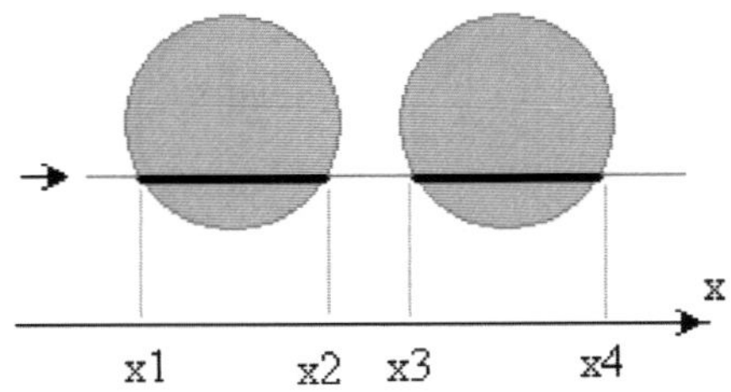

Figure 3. Set of points and normals

When analyzing interchange order for outer and back faces, we can reveal embedded surfaces or intersected surfaces of the geometrical model. Based on these results, we can formulate a condition: geometrical model contains no gross routing errors, if the ordered set of intersection points and appropriate normals contains first outer face, last back face, and the quantities of outer and back faces match.

For example, for embedded or intersected objects (Figure 4), order of faces will be as follows:

$\{x_1, outer\}, \{x_2, outer\}, \{x_3, back\}, \{x_4, back\}$.

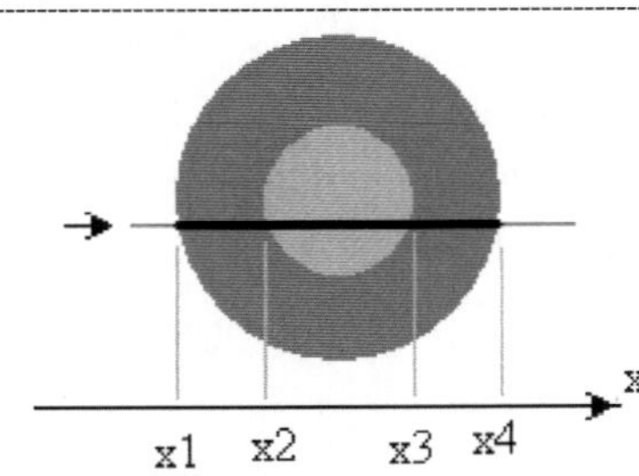

Figure 4. Embedded objects.

If the numbers of outer and back faces do not coincide, it is due to incomplete surfaces. For example, the following set (Figure 5) shows us errors in model geometry, which result in cavities in the end product.

$\{x_1, outer\}$.

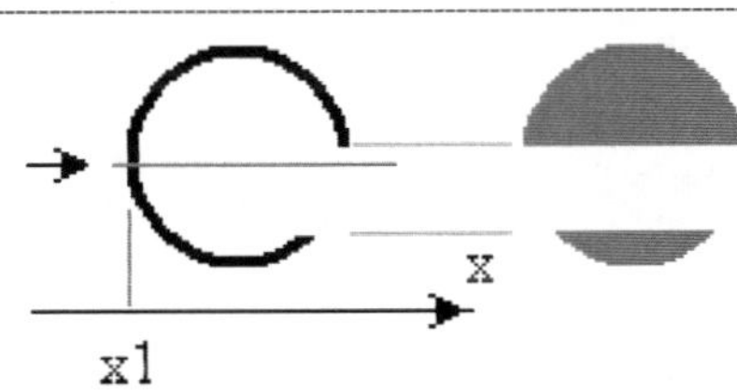

Figure 5. Model with geometry and layer-by-layer errors

Such errors can be corrected using a complicated routing method algorithm. If we save the history of previous routing results and restore the data, we will avoid the emptiness in presentation of surfaces and improve layer-by-layer model presentation. For example, if we save previous scanning data when the scanning vector goes bottom-up, we will be able to complete the surface with the emptiness error (Figure 6).

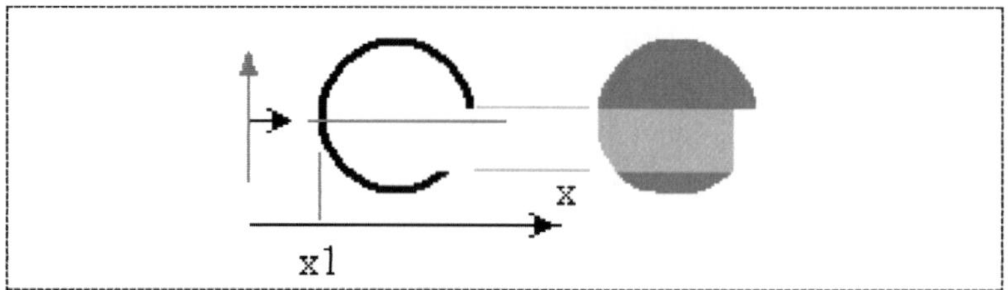

Figure 6. Restoration of model with emptiness error

The developed method is realized in the Program for Additive Manufacture Preparation (PAMP) [4]. Based on discrete model presentation in STL format, layer-by-layer model presentation is realized with necessary precision. Generated layers can be exported in raster format (BMP files) or can be used for G-code generation.

PAMP analyzes routing results for layer-by-layer presentation and checks for model geometry errors. Then, found geometry errors can be corrected automatically using the developed algorithm (Figures 7a and 7b). In Figure 7a, the geometry error and its layer position in the graph is marked red.

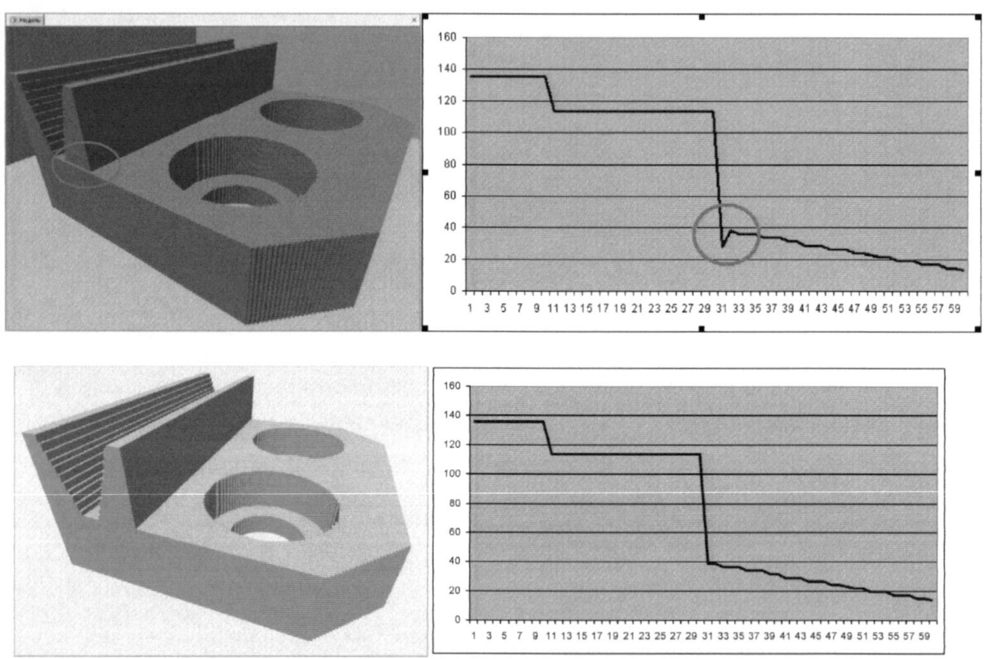

a)

b)

Figure 7. Correction of 3D-model geometry error: a) layer-by-layer representation of a 3D-model with geometry error (on the right is the dependence of layer volume on height); b) corrected geometry of the model (on the right is a new dependence of layer volume on height)

4. Development of the multiprocessing algorithm for batch file forming

Every additive manufacturing process is based on batch file, which is to be executed by 3D-printing equipment. The batch file contains commands for the movement of a printing jet or laser beam (depends on 3D-printing technology).

The role of the batch file in additive manufacturing is shown in Figure 8.

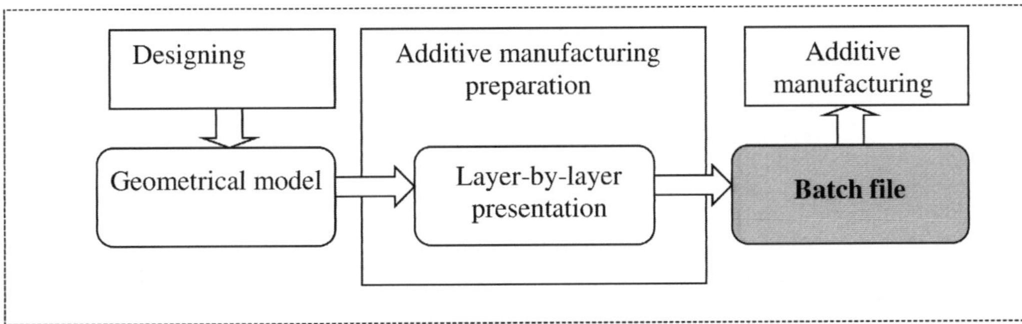

Figure 8. Batch files in additive manufacturing

Most 3D-printers use standard G-Code format as the batch file format. The batch file in G-code format is a text file, which contains setting commands and ordered commands of blank cycle (G0) and working stroke (G1). Example: G1 X48.700 Y1.049 E2.31026 F1350.000.
Geometrical model (a), layer-by-layer presentation (b) and visual presentation of the batch file (c) are shown in Figure 9.

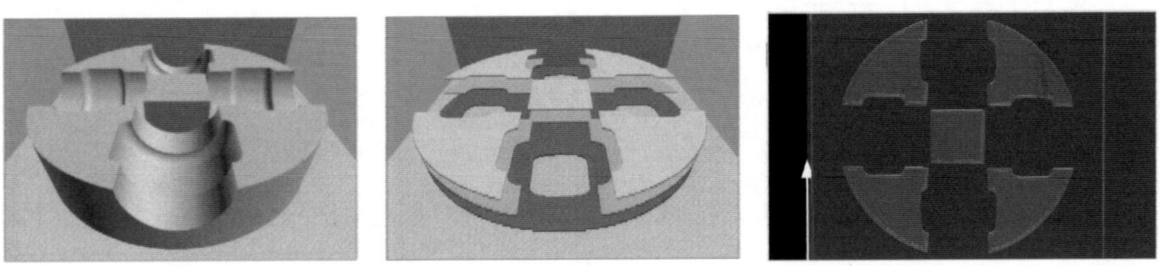

(a) (b) (c)

Figure 9. Model presentation stages: (a) geometrical model; (b) layer-by-layer presentation; (c) visual presentation of the batch file

Input data for batch file generation contain layer-by-layer model presentation. Layer can be presented as a set of parametric contours or as a raster BMP file.
Command presentation has two sections: perimeter and filling. It is necessary for edge smoothness and for interior content.
Perimeter section requires continuous closed contour bypass. For filling section, it is necessary to use closed contour filling. The presentation algorithms are shown in Table 1.

Table 1. Presentation algorithms

Input data	Result	Algorithm type
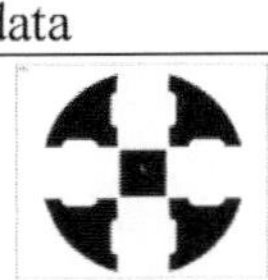 Plane layer	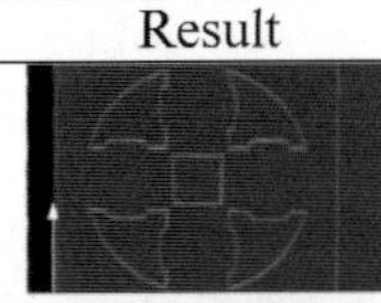 Command presentation of perimeter	Continuous closed contour bypass
 Plane layer	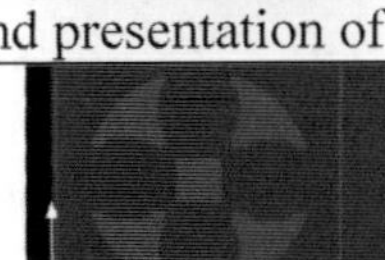 Command presentation of filling	Continuous closed contour filling

The authors developed algorithms for batch files in G-Code format (PAMP). Developed software components were adapted for selective laser sintering.

The scheme for paralleling batch files is presented in Figure 10.

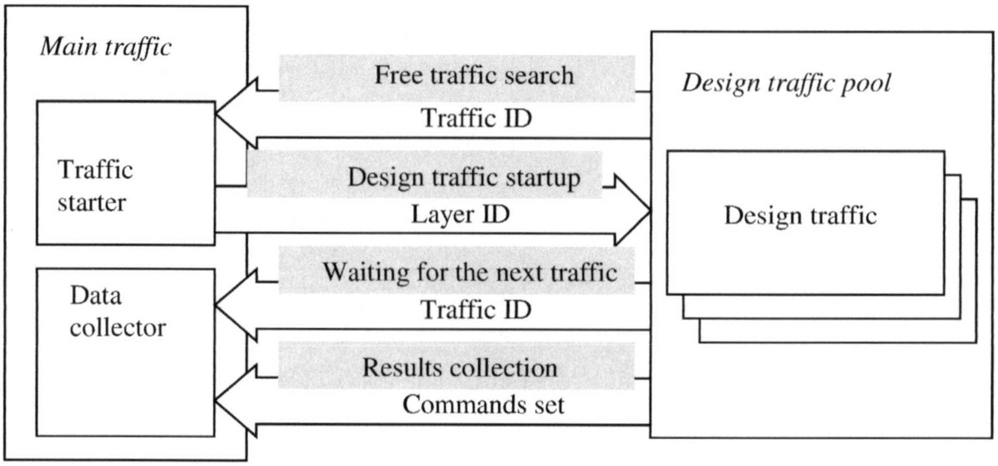

Figure 10. Batch files paralleling scheme

5. Conclusion

In Moscow Aviation Institute, new software for additive manufacturing preparation was developed—PAMP—with funding from the Russian Federal Targeted Program "Research and Development in Priority Fields of the Scientific and Technological Complex of Russia for 2014-2020", within the framework of arrangement no. 2.2, agreement no. 14.586.21.0019. PAMP is designed for G-code generation for additive manufacturing equipment and can be used for Selective Laser Sintering (SLS) and Fused Deposition Modelling (FDM). The main benefits of the developed software are:

1) PAMP allows user to check 3D-model geometry for errors and correct them before additive manufacturing preparation;
2) it is based on multiprocessing algorithms, which allows to speed up the additive manufacturing preparation process.

PAMP was tested during realization of real additive manufacturing tasks using FDM-technology and it have demonstrated good results: preparation of tested models for 3D-printing was successful, errors in 3D-models geometry were found and corrected, quality of produced models using PAMP-generated G-code was acceptable. Generation of G-code using developed multiprocessing algorithm was faster than that using single-processor algorithm according to linear relation.

References

[1] Dongdong Gu 2015 *Laser additive manufacturing of high-performance materials* (Berlin: Springer)

[2] Gibson, Rosen, Stucker 2010 *Additive manufacturing technologies.* (Berlin: Springer)

.

[3] Patri K. Vevuvinod, Weiyin Ma 2004 *Rapid prototyping. Laser-based and other technologies* (Berlin: Springer)

[4] Anamova R, Avtushenko A, Ivanov A, Ripetskiy A, Osipov A 2015 *J.Vestnik BGTU.* **2** (46) 8

IOP Conf. Series: Materials Science and Engineering **140** (2016) 012004 doi:10.1088/1757-899X/140/1/012004

Development of bioceramic material for spinal surgery implants

T Sablina[1,a], N Savchenko[1,2,3], A Pshenichnyy[2], M Grigoriev[1,2,3], S Buyakova[1, 2, 3] and S Kulkov[1, 2, 3]

[1]Institute of Strength Physics and Materials Science of the Siberian Branch of the Russian Academy of Sciences, 2/4, pr. Akademicheskii, Tomsk, 634055, Russia,
[2]National Research Tomsk State University, 36, Lenin Avenue, Tomsk, 634050, Russia,
[3]National Research Tomsk State Polytechnic University, 30, Lenin Avenue, Tomsk, 634050, Russia

[a] Corresponding author: *sabtat@ispms.tsc.ru*

Abstract. Highly porous zirconia-based ceramics were prepared. The ceramic samples sintered at 1600°C had porosities from 40% to 43%, with pore size ranges as follows: "big pore" 100-220 μm and "small pore" 0.8-8 μm. This makes the ceramic structure to be very similar to the structure of the natural spinal bone. The level of mechanical properties of the synthesized zirconia-based ceramics is determined by the pore sizes. The values of the compressive strength and the effective Young's modulus are very similar to those characteristics of the natural spinal bone.

1. Introduction

Additive technologies are the most dynamically developing directions of bio-implants production [1]. Analysis of the publications as well as current trends of research in the field of spinal prosthesis showed that until recently, that the largest part of the intervertebral disc prosthesis made of metal, the main advantage is ease of manufacturing and processing [2-8].

An analysis of publications in the field of spinal prosthesis showed that until recently, the intervertebral disc prostheses have been made of metal [2-8]. However, an adverse effect of metal on the adjacent tissue and the whole organism forced searching new materials for spinal implants.

An alternative to metals in the manufacture of implants are ceramic materials. Ceramic materials are closest to the bone material in terms of biochemical compatibility.

Zirconia-based bioceramics produced from nanopowders is considered as one of the most promising materials for additive biotechnologies due to possessing a unique combination and range of properties including strength and fracture toughness, hardness and wear resistance, biocompatibility, bacteriostatic, and osteointegration [2].

The aim of this paper is to study the physical and mechanical properties of the sintered porous zirconia ceramics.

2. Experimental materials and investigation procedure

The Zr, Mg, and Al aqueous salt solutions served as the original materials for obtaining porous ceramic ceramics; the salts were converted into powders in the high-frequency plasma discharge. The powders had the composition as follows: ZrO_2 +9 mol % MgO. Samples for sintering were prepared by slip casting. The required porosity and configuration of the pores were provided by introducing in a ceramic mass a predetermined amount of organic binder and blowing agent such as granular UHMWPE powder with grain size in the range 90 to 200 μm. The sintering of the ceramic samples was performed in an air furnace at 1600°C, isothermal holding at the sintering temperature was 1, 5, 10 hours. Investigation of the structure of sintered ceramics was carried out by optical microscope "Altami M1" and scanning electron microscope "VEGA3 TESCAN". X-ray diffraction data were obtained using an X-ray diffractometer with CuK_α radiation with a step of 0.05° in the range of 2 theta angles from 20 to 80 degrees. Samples of porous ceramics were uniaxial compressed using a universal testing machine "GP DLC 30 kN". Effective Young's modulus (E) was determined as the angle of a tangent line slope to a straight-line segment of the "stress-strain" curve.

3. Results and discussion

Studying the structure of sintered ceramics showed that porosity reduced for increasing the sintering time from 1 up to 10 hours (Figure 1a). The ceramic samples had porosities with pore sizes in two categories: "big pores" around 100-220 μm and "small pores", around 0.8-8 μm (Figure 1).

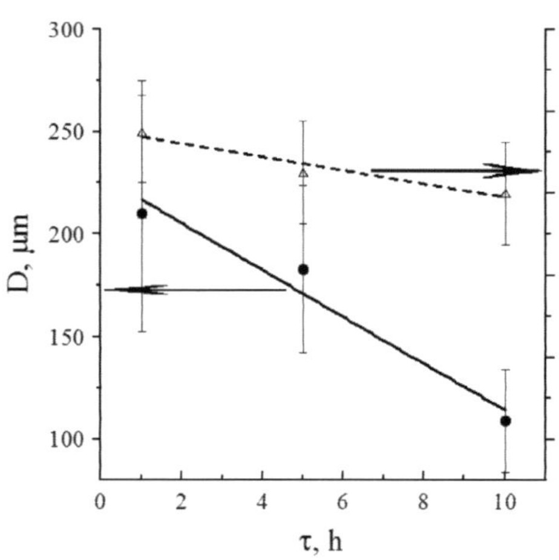

(a)

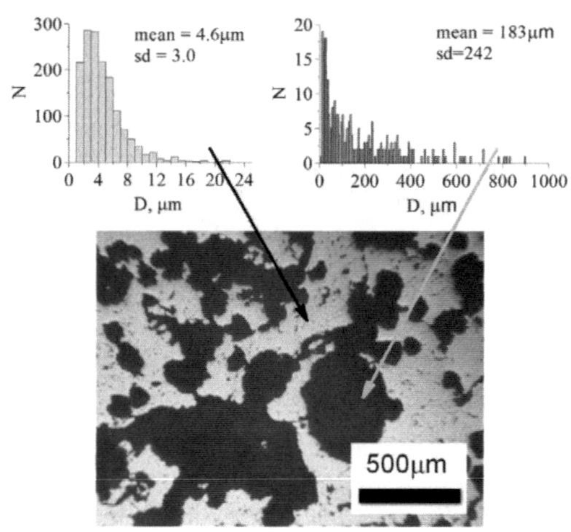

(b)

Figure 1. Dependence of porosity (P) and the average pore size (D) of the isothermal holding time (a), typical structure and distribution of pores in ceramics, sintered at a temperature of 1600°C, for 5 h (b).

The average "small pore" sizes decreased from 8.5 to 0.8 μm, while sintering time increased from 1 to 10 hours.

The average "big pores" sizes decreased from 210 to 110 μm when sintering time increased from 1 to 10 hours (Figure 1a). The formation of "Big pores" is associated with blowing agent removal during sintering of ceramics.

The fact that D the average pore sizes of the two types decreased with increasing the holding time of sintering may be associated with the densification and recrystallization ceramics during sintering.

Figure 2 shows the scanning electron microscope images of porous ceramics fracture surfaces. The fracture of porous ceramics is mainly by intercrystalline mechanism with microcracks shown by arrows in Figure 2b. The analysis of the human bone porous structure showed that it is inhomogeneous by the bone depth [5]. A spinal bone consists of cortical (compact) and trabecular (spongy) substances. They both form a lamellar bone tissue. The pore size is depending on the area of bone and ranges from

0.5 to 500 μm. Comparative studies of the structure of the porous ceramic materials and human trabecular bone [5] showed the similarity between structural components of porous ceramics and with the bone tissue.

Figure 3 shows the XRD patterns of porous ceramics after sintering at 1600° C for 1 and 10 h. As can be seen from this Figure the ceramics is composed of two phases: monoclinic (M) ZrO2 phase and cubic (K) ZrO_2 phase. Quantitative phase analysis showed that cubic/monoclinic phase ratio is 6/4 and corresponds to that of known ZrO_2 - MgO state diagram [9]. Increasing the sintering time from 1 to 10 hours leads to reducing the cubic phase content.

Figure 4 shows stress–strain curves of the porous ceramic sample and human trabecular bone [10] after uniaxial compression tests. This kind of deformation curve is typical for the fracture of porous zirconia-base ceramics [11]. However, a distinguishing feature of the deformation curves obtained for porous zirconia-base ceramics is that there is a deviation from linearity at high stress as well as downward oscillating branch with a gradual decrease in the stress due to the accumulation of microcracks (Figure 2b).

Ceramic mechanical properties studies have shown that both compressive strength and effective elasticity modulus increased when decreasing the pore sizes. The compressive strength for ~40% porosity materials is 22 MPa. Effective elasticity modulus determined from the slope of the deformation curves for the studied materials is in the range 1.0-1.8 GPa. Quasi-static compression tests on porous ceramics showed that the material had satisfactory strength for the use as osteogenesis material.

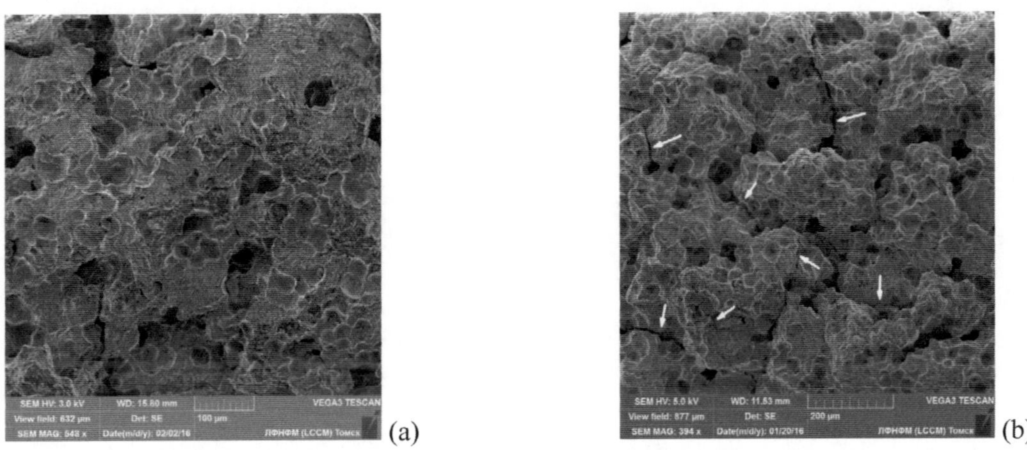

Figure 2. The SEM image of the fracture surface of porous ceramics (after the quasi-static compression tests). Arrows in the Figure 2b, indicate of microcracks.

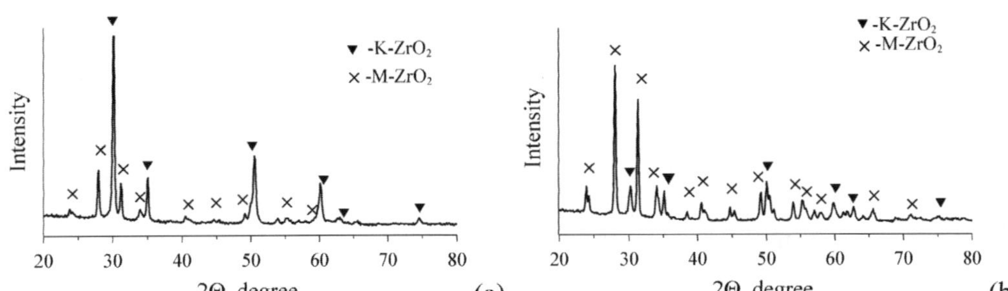

Figure 3. The XRD pattern of polished porous ceramics surface after sintering 1 h (a) and 10 h (b) at 1600°C.

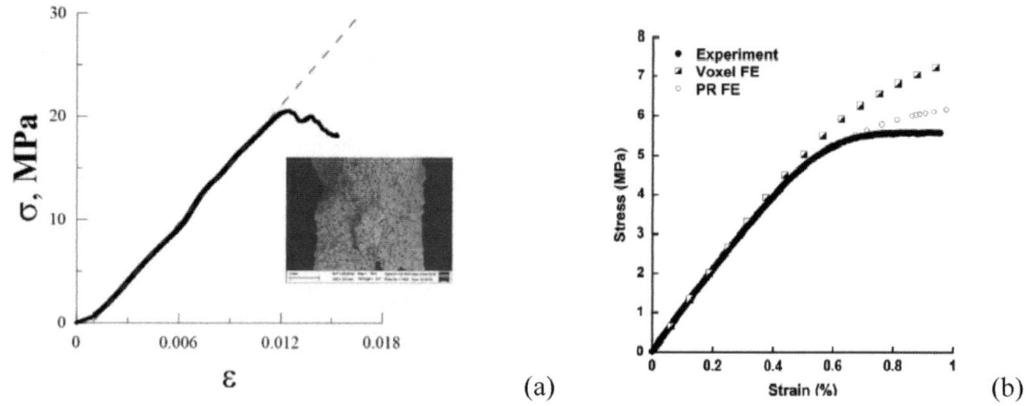

Figure 4. Stress–strain curve and macrograph of ceramic with a porosity of 40% subjected to compression (a); the stress–strain curve of human trabecular spinal bone subjected to compression (b) [10].

4. Conclusion

It was shown that after sintering at 1600°C that zirconia ceramic structure was characterized by the presence of pores with pore sizes in two categories such as "big pore" 100-220 μm and "small pore" 0.8-8 μm. This makes the ceramic structure very similar to the structure of the natural spinal bone.
The level of mechanical properties of the synthesized zirconia-based ceramics is determined by the pore size. The values of the compressive strength and the effective Young's modulus are very similar to those of the natural spinal bone.

Acknowledgement
This work was partially financially supported by the Ministry of Education and Science of the Russian Federation in the framework of Agreement No. 14.607.2 1.0069 (RFMEFI60714X0069).

References

[1] Bartolo P, Kruth J P, Silva J, Levy G, Malshe A, et al. 2012 *CIRP Annals - Manufacturing Technology* **61** 635–655
[2] McEntire B J, Bal B S., Rahaman M N., Chevalier, Pezzotti G 2015 *Journal of the European Ceramic Society* **35** 4327–4369
[3] Davis W, Allouni A K, Mankad K, Prezzi D, Elias T, Rankine J, Davagnanam I 2013 *Clinical Radiology* **68** 64-74
[4] Staiger M P, Pietak A M, Huadmai J, Dias G 2006 *Biomaterials* **27** 1728–1734
[5] Wua S, Liu X, Yeung K W K, Liu C, Yang X 2014 *Materials Science and Engineering* **R 80** 1–36
[6] Savchenko N L, Sevostyanova I N, Sablina T Y, Molnar L, Geber R, Gomze L A, Kulkov S N, Gomze L N 2014 *Építôanyag - Journal of Silicate Based and Composite Materials* **66** 44-47
[7] Savchenko N L, Sevostyanova I N, Sablina T Yu, Gomze L, Kulkov S N 2014 *Physical Mesomechanics of Multilevel Systems 2014: AIP Conf. Proc* **1623** 547-550
[8] Grigoriev M V, Kulkov S N 2011 *Russian Physics Journal* **1** 1305–1311
[9] Odnik H M, McMurdie H F Phase Diagrams for Zirconium and Zirconium system. 1999 *Maryland: Phase Diagram Data Center*
[10] Ji W, Bin Z, Sherry L, Aaron J. F, Arnav S, Xiutao S, Adams M, Keaveny T M, Guo X E 2015 *Bone* **72** 71–80.
[11] Savchenko N L, Sablina T Yu, Sevostyanova I N, Buyakova S P, Kulkov S N 2016 *Russian Physics Journal* **58**/11 (*Russian Original* 2015 11 56-60).

The effect of hydrogen on strain hardening and fracture mechanism of high-nitrogen austenitic steel

G G Maier[1], E G Astafurova[1], E V Melnikov[1], V A Moskvina[2], V F Vojtsik[2], N K Galchenko[1] and G N Zakharov[3]
[1]Institute of Strength Physics and Materials Science, Siberian Branch of Russian Academy of Sciences, Akademichesky Avenue 2/4, Tomsk, 634055, Russia
[2]National Research Tomsk Polytechnic University, Lenina Avenue 30, Tomsk, 634050, Russia
[3]National Research Tomsk State University, Lenina Avenue 36, Tomsk, 634050, Russia

E-mail: elena.g.astafurova@gmail.com

Abstract. High-nitrogen austenitic steels are perspective materials for an electron-beam welding and for producing of wear-resistant coatings, which can be used for application in aggressive atmospheres. The tensile behavior and fracture mechanism of high-nitrogen austenitic steel Fe-20Cr-22Mn-1.5V-0.2C-0.6N (in wt.%) after electrochemical hydrogen charging for 2, 10 and 40 hours have been investigated. Hydrogenation of steel provides a loss of yield strength, uniform elongation and tensile strength. The degradation of tensile properties becomes stronger with increase in charging duration – it occurs more intensive in specimens hydrogenated for 40 hours as compared to ones charged for 2–10 hours. Fracture analysis reveals a hydrogen-induced formation of brittle surface layers up to 6 μm thick after 40 hours of saturation. Hydrogenation changes fracture mode of steel from mixed intergranular-transgranular to mainly transgranular one.

1. Introduction

Austenitic stainless steels are frequently used materials for industrial applications due to their high ductility at cryogenic temperatures and low aggressive environment embrittlement. It concerns mainly AISI 300-series Cr-Ni steels [1-3]. However, the presence of expensive alloying elements – nickel – and low yield stresses limit their use. The requirements of low costs have generated an interest in replacement of nickel and use nitrogen and manganese as alloying elements. The high-nitrogen Cr-Mn steels have good mechanical properties and corrosion resistance and could be of great interest for a wide range of industrial application – in particular, for producing wear-resistant coatings or items using electron-beam welding [4] or other methods of additive technologies. An indubitable advantage of such materials could be their good strength to wear combined with resistivity to degradation in aggressive atmospheres, for instance, in hydrogen.

Hydrogen embrittlement behavior of high-nitrogen steel is defined by the complex of several parameters; among them are stacking fault energy (SFE), steel composition, regime of hydrogenation and hydrogen content. Michler et al. [5] demonstrated that replacement of Ni on Mn and N of Cr-Mn austenitic steel decrease the resistance to hydrogen embrittlement. Authors [6] reported that amount

Published under licence by IOP Publishing Ltd

of nitrogen in solid solution of Cr-Mn steel had strong effect on its ductility under hydrogen charging: specimens with highest nitrogen content (0.6N, wt.%) dissolved more hydrogen and had a higher loss in ductility in comparison with 0.3N wt.% specimens. At the same time, Stoltz and co-authors [7] demonstrated that nitrogen in Cr-Ni-Mn-N steel lowered the SFE value until the N concentration reached 0.24 wt.%, but further increase in nitrogen content up to 0.52 wt.% did not change the value of SFE. Michler et al. [8] also found that SFE, as well as martensitic transformations, did not directly control the susceptibility to hydrogen-enhanced embrittlement of austenitic steels, but initial deformation mode was one important parameter respected for tensile ductility in hydrogen atmosphere. Igata et al. [9] investigated dislocation structure and ductility of Fe-Cr-Ni-Mn-0.35N, wt.% austenitic steel after hydrogenation. They reported that hydrogen decreased SFE, suppressed cross slip, decreased the dislocation multiplication rate and made ductility lower [9]. The effect of hydrogen on the properties of high-nitrogen steels was not studied in detail, and the problem of hydrogen environment embrittlement is still far from a final decision.

The present study tried to shed light on tensile behavior and fracture mechanism of high-nitrogen austenitic steel after hydrogenation.

2. Materials and methods

A high-nitrogen Fe-20Cr-22Mn-1.5V-0.2C-0.6N, wt. % steel was selected as an object of investigation. Steel specimens were quenched after 1100°C, 1 h. into cold water.

Before hydrogenation the specimen were cut dog-bone shaped flat tensile specimens with nominal dimensions of 18 mm x 2.7 mm x 0.7 mm (thickness) in the gauge section. Mechanical grinding and a final electrochemical polishing (50 g CrO_3, in 200 g H_3PO_4) were employed to remove the entire processing-affected surface layer. The final thickness of tensile specimens was 0.5 mm.

Electrochemical hydrogen charging was performed at current density of 50 mA/cm^2 for t_{hyd} = 2, 10, 40 hours at room temperature in 1M H_2SO_4 water solution, containing 0.25g l^{-1} of CH_4N_2S as a recombination poison. After hydrogenation, the tensile tests were conducted at room temperature and a strain rate of 5.6×10^{-4} s^{-1} using Instron 3369 electromechanical machine.

The fracture and side surface of specimens were examined using a LEO EVO 50 (Zeiss, Germany) scanning electron microscope (SEM).

3. Results and discussion

Figure 1 shows the tensile stress-strain curves for hydrogen-free and hydrogen-charged specimens. The detailed results of tensile tests are summarized in Table 1.

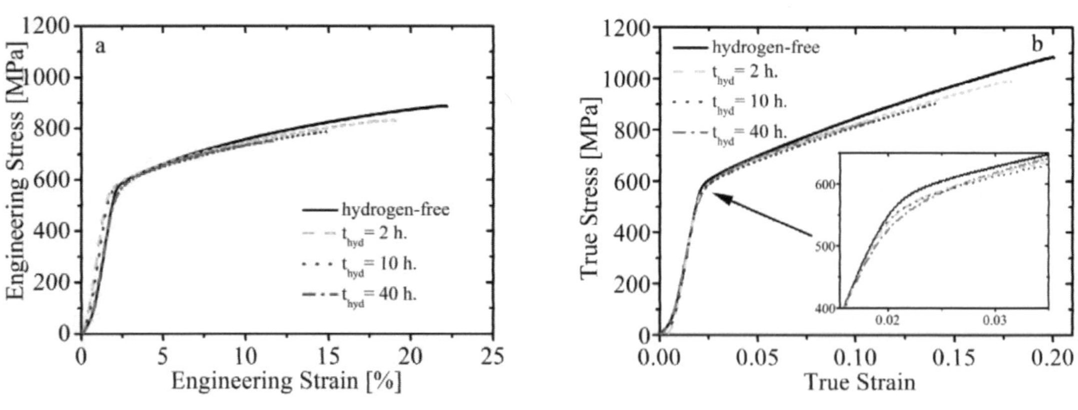

Figure 1. Stress-strain curves depending on hydrogenation duration: (a) engineering stress versus engineering strain; (b) true stress versus true strain.

Table 1. Influence of hydrogenation regime on yield stress ($\sigma_{0.2}$), uniform elongation (ε_{un}), tensile strength (σ_B), strain-hardening coefficient $\theta = (\sigma_{un} - \sigma_0)/(\varepsilon_{un} - \varepsilon_0)$ and hydrogen embrittlement index (E_H) according to true stress-true strain curves.

Treatment	$\sigma_{0.2}$ (MPa)	ε_{un} (%)	σ_B (MPa)	Θ (MPa)	E_H (%)
Hydrogen-free	586	20	1084	2800	0
t_{hyd} = 2 h.	538	17	989	2770	15
t_{hyd} = 10 h.	541	14	902	2780	30
t_{hyd} = 40 h.	520	11	838	3140	45

Increase in hydrogenation duration slightly decreases a yield stress of steel: from 586 MPa in hydrogen-free down to 520 MPa in hydrogen-charged specimens (for 40 h.). The elongation deteriorates under hydrogen saturation from 20 % for hydrogen-free specimens to 17 %, 14% and 11% after hydrogen charging for 2, 10 and 40 hours respectively. The hydrogen embrittlement index (E_H), representative of % loss in uniform elongation due to hydrogen alloying, was calculated using the following equation (Table 1):

$$E_H = [(\varepsilon_{un0} - \varepsilon_{unH}) / \varepsilon_{un0}] \cdot 100\%,$$

where ε_{un0} and ε_{unH} are values of uniform elongation for hydrogen-free and hydrogen-charged specimens respectively. Analysis of data in table 1 shows that if the duration of hydrogenation reaches 40 hours, there is a maximum loss of uniform elongation and the highest value of E_H (45%) is observed. At the same time, the value of strain hardening coefficient does not change sufficiently with increasing of hydrogenation duration.

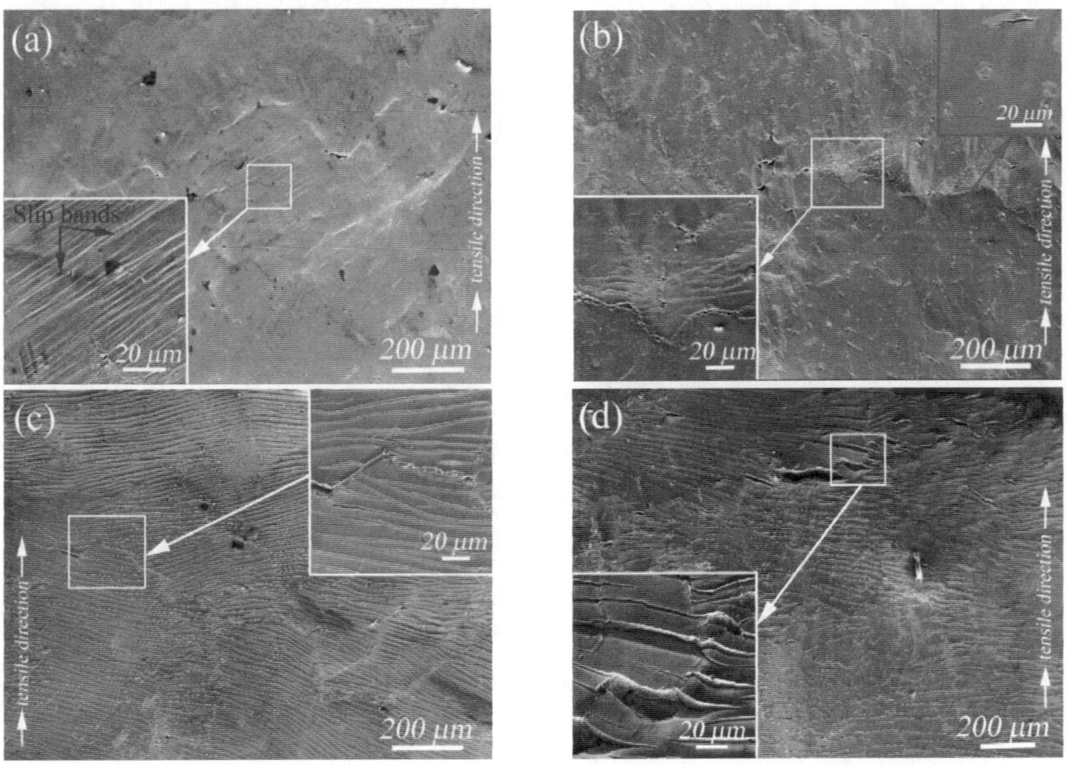

Figure 2. SEM micrographs of side surfaces of specimens without hydrogenation (a); after hydrogenation for 2 h. (b); after hydrogenation for 10 h. (c); after hydrogenation for 40 h. (d).

The micrographs of the side surfaces of specimens in initial state and after hydrogenation are presented in figure 2. The hydrogen-free specimen demonstrated slip bands inside of grains with the step of about 3 μm (Figure 2a). After 2 hours of hydrogen charging and tension, hydrogen-induced cracks are observed along grain boundaries and inside of them (Figure 2b). Some of these cracks with a length of 5-20 μm are detected inside of grains; other long cracks propagates through the grains perpendicular to tensile direction and have distance of 4 μm between them (Figure 2b, inserts). The micrographs clearly show that an increase in hydrogenation duration increases the distance between cracks and their depth: to 12 μm after 10 hours and to 17 μm after 40 hours of hydrogen saturation (Figures 2c-d). These differences in distance between cracks and crack depth are associated with the different thickness of hardened (brittle) layer and with different tensile elongation of specimens to fracture (Table 1).

The fracture surfaces of all specimens were examined to observe the influence of hydrogen on fracture behavior of the high-nitrogen steel (Figure 3). The hydrogen-free specimens were found to exhibit a predominantly intergranular ductile fracture (Figure 3a), but regions with transgranular fracture was also observed. A main feature of the fracture surface of hydrogen charged samples in comparison with hydrogen-free one is an appearance of brittle surface layer on the specimen surface (Figures 3 b-d). In the central area, the fracture morphology of the hydrogen-charged specimens is ductile. It contains regions with coarse dimples and fine dimples with parallel edges closely spaced to each others. Rough secondary macrocracks is observed in central part of hydrogenated specimens.

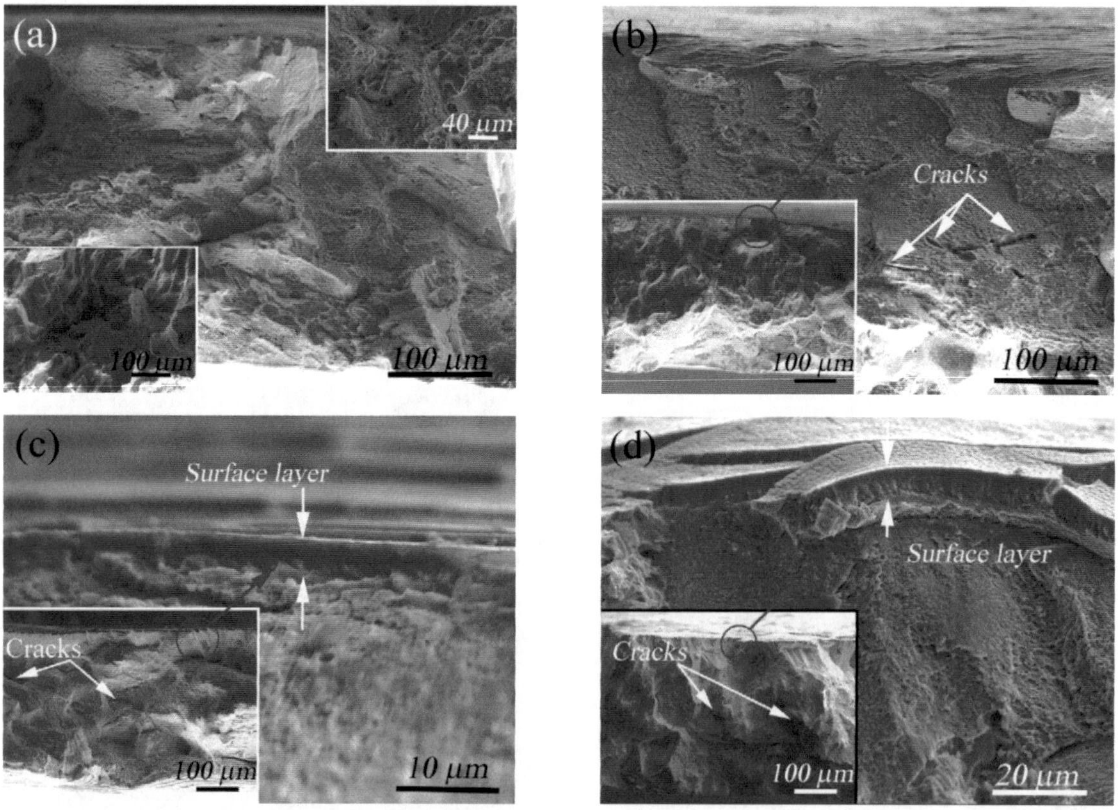

Figure 3. SEM micrographs of the fracture surfaces of hydrogen-free (a) and hydrogen-charged for 2h. (b), 10 h. (c), 40 h. (d) specimens.

The thin surface layer with the mean thickness of 1.5 μm forms after 2 h. hydrogenation of steel (Figure 3 b). The hydrogenation for 10 and 40 hours provides thicker brittle surface layers with a thickness of 3.5 μm and 6 μm respectively (Figure 3 c, d). The hydrogen-assisted surfaces layers consist of two regions: the first one (surface) is homogeneous and brittle (cleavage-like); the second one (subsurface) characterized by columnar structure with the elongated dimples oriented perpendicular to surface. The fracture mode in hydrogen charged specimens is also mixed, but predominantly transgranular with parallel elongated dimples in different directions. The fracture character of the specimens after hydrogenation for 40 hours does not significantly change in comparison with the specimens charged for 10 hours (Figures 3 c-d). The embrittlement of a surface layer and nucleation of secondary macrocracks and microcracks in central ductile part of the hydrogenated steel specimens provoke their fast failure compared to hydrogen-free one.

The results clearly show that hydrogen charging changes fracture regime of high-nitrogen steel and provides the significant loss in uniform elongation of specimens. The change in fracture regime and appearance of parallel elongated dimples is connected with the fact that hydrogen reduces stacking fault energy of austenite, enhances slip localization and increases the nucleation of voids at the intersection of slip bands. Since deformation bands are usually planar, have regular spacing and intersect along lines, such nucleating voids appear as parallel elongated dimples. Similar feature was demonstrated for Cr-Ni austenitic steels by Nibur et al. [10]. Michler et al. [8] reported that high concentration of interstitial atoms in steel can influence the deformation mode not only via SFE but also by effects like short-range ordering and microtwinning. These effects, obviously, could arise in our case due to high concentration of nitrogen in solid solution and pronounced mechanical twinning in high-nitrogen steels [11]. The observed cracks in hydrogen-charged specimens can be also attributed to the hydrogen-enhanced localized plasticity [12] in austenite. Another possible explanation of change of fracture mode and appearance of cracks in hydrogen-charged specimens connected with two-phase composition (austenite and small portion of ferrite) of steel investigated. The diffusivity of hydrogen in ferrite higher than in austenite at room temperature [13], that could provoke an earlier fracture in ferrite phase compared to austenite one. The hydrogen can assisted fracture due to accumulation of hydrogen in ferrite-austenite interfaces and leads to embrittlement of structure. Obtained experimental results tend to support all presented explanations and required further detailed investigation of the steel structure.

4. Summary
The effect of hydrogen on tensile properties and fracture behavior of austenitic steel Fe-20Cr-22Mn-1.5V-0.2C-0.6N (in wt.%) has been investigated. Hydrogen decreases yield strength, uniform elongation and tensile strength of steel with increasing of hydrogenation duration. The formation of brittle surface layers occurs under hydrogenation, the mean thicknesses of these layers depends on charging regime and reaches 6 μm after hydrogenation for 40 hours. Surface layers crack under tension in cleavage-like manner. Analysis of fracture revealed that hydrogenation leads to change in fracture mode from mixed intergranular-transgranular to mainly transgranular one and promoted secondary cracking inside of specimens.

Acknowledgments
This research was partly supported by Russian Foundation for Basic Researches (project No. 15-38-20056, in part of mechanical characterization) and Russian President Scholarship (SP-160.2016.1, in part of electron microscopy). The experimental results were obtained using equipment of Center «Nanotech» of Institute of Strength Physics and Materials Science, Siberian Branch of Russian Academy of Sciences.

References
[1] Michler T, Naumann J, Hock M, Berreth K, Balogh M P, Sattler E 2015 *Mat. Sci. and Eng. A* **628** 252–61

[2] San Marchi C, Michler T, Nibur K A and Somerday B P 2010 *Int. J. Hydrogen Energy* **35 (18)** 9736–45

[3] Zhang L, Wen M, Imade M, Fukuyama S and Yokogawa K 2008 *Acta Mater.* **56 (14)** 3414–21

[4] Narkevich N A, Durakov V G , Tagiltseva D N, Shulepov A I and Smirnov A I 2013 *Phys. Met. Metall.* 2013 **114(6)** 535-44

[5] Michler T, Naumann J 2010 *Int. J. of Hydrogen Energy 35* 1485–92

[6] Phaniraj M P, Kim H-J, Suh J-Y, Shim J-H, Park S-J and Lee T-H 2015 *Int. J. of Hydrogen Energy* **40** 13635–42

[7] Stoltz R E and Vandersande J B *Metall. Trans A* 1980 **11A** 1033–37

[8] Michler T, San Marchi C., Naumann J, Weber S, Martin M 2012 *Int. J. of Hydrogen Energy* **37** 16231–46

[9] Igata N, Fujiga T and Yumoto H 1991 *J. Nucl. Mater.* **179** 656-58

[10] Nibur K A, Somerday B P, Balch D K, Marchi C S 2009 *Acta Mater.* **57** 3795–3809

[11] Gavrilyuk V G, Berns H 1999 High nitrogen steel Berlin (Springer Verlag Publ.) p 378

[12] Birnbaum H K and Sofronis P 1994 Mat. Sci. and Eng. A **176** 191–202

[13] Younes C M, Steele A M, Nicholson J A and Barnett C 2013 Int. J. Hydrogen Energy 38 4864–76

The influence of ZrB_2-SiC powders mechanical treatment on the structure of sintered ceramic composites

S Buyakova[1,2,3], A Burlachenko[1,3], Yu Mirovoi[1], I Sevostiyanova[1], S Kulkov[1,2,3]

[1]Institute of Strength Physics and Materials Science SB RAS, 2/4, pr. Akademicheskii, Tomsk, 634055, Russia
[2]Tomsk Polytechnic University, 30, Lenin Avenue, Tomsk, 634050, Russia
[3]Tomsk State University, Tomsk, 36, Lenin Avenue, Tomsk, 634050, Russia

Corresponding author: *y.a.mirovoy@gmail.com*

Abstract. The effect of mechanical treatment by planetary ball milling on the properties of hot pressed ZrB_2 - SiC ceramics was studied. It was shown that material densification after mechanical treatment is finished at initial stages of sintering process. Addition of SiC leads to an essential increase of sample density to 99% of theoretically achievable for powder with 2% of SiC, as compared with ZrB_2 with the density less than 76%. It was demonstrated that all defects that were accumulated during mechanical treatment are annealed during hot pressing, and there are no changes of CDD values in sintered ceramics.

1. Introduction

The refractory compounds are the basic components of materials used in high-temperature engineering, such as thermal protection of space vehicles, electronics, etc. Among materials with high melting temperature, a special attention is paid to ceramic ZrB_2-based composites; however, strong covalent bonds are the reason of low diffusion mobility of an atom, which appreciably impedes the manufacturing of products.

It is known [1], that addition of SiC to ZrB_2 increases the density of sintered materials due to smaller melting temperature of SiC. Besides, ZrB_2-SiC ceramics are characterized by high thermochemical stability, including high stability to oxidation in conditions of super-high temperatures.

The ZrB_2-SiC composites are usually obtained by sintering powders under pressing at temperatures higher than 2000 °C [2], and to lower the sintering temperature, the powders undergo mechanical treatment in high-energy (planetary) mills. In this case, subsequent sintering will be activated due to increased number of defects, acceleration of diffusion processes and simplification of plastic flows during sintering [3], so sintering can be carried out under pressing, i.e. it is possible to realize hot pressing or SPS-process [4].

Unfortunately, data on the influence of mechanical treatment on the properties of powders and the process of subsequent hot pressing are poorly investigated. Therefore, the main aim of this paper is to study the influence of mechanical treatment of ZrB_2-SiC powders on their properties and properties of ceramic composites sintered by hot pressing.

2. Materials and Experimental Procedure

The research was carried out using powder mixtures of ZrB_2 (d_{50} = 2.5 μm) and SiC (d_{50} = 4.2 μm) with SiC content of 10, 15 and 20 vol.%. The powders were mechanically treated in a planetary mill at the centrifugal acceleration of approximately 30g. The duration of processing was up to 20 minutes.

Hot pressing of ceramic composites was carried out at the temperature of 1800 °C and pressure of 50 MPa with isothermal sintering for 30 minutes. According to [5], X-ray with CuKα source was used to measure raw density of mixtures, phases, structure and coherently diffracting domains (CDD). Scanning electron microscope observations operated at 20-30 kV were used to determine the structure and average grain size using Tescan VEGA-3SBH.

3. Results and Discussion

Figure 1 shows the dependence of relative density ρ_r/ρ_t (ρ_r raw density, ρ_t theoretical density) vs. treatment time. Evidently, for any powder, the increase of treatment time was accompanied by increase of its raw density; in addition, the morphology of particles has appreciably changed, from separate particles in the beginning state up to the formation of agglomerates in the end of treatment (Figure 2). Before the treatment, ZrB_2 powder had the smallest raw density; however, after processing, its density has increased essentially and exceeded the raw densities of powder mixtures with SiC. These mixtures in initial state had very close values ρ_r/ρ_t, but after the treatment, the powder with the maximal content of SiC had higher raw density. The X-ray phase analysis of mixtures has shown that during the treatment there were no changes; addition of SiC to the mixture leads to the occurrence of its peaks. With increased treatment time, we have found a broadening of peaks due to increasing number of lattice defects and decrease of CDD or grain size from 46 down to 37 nm (Figure 3).

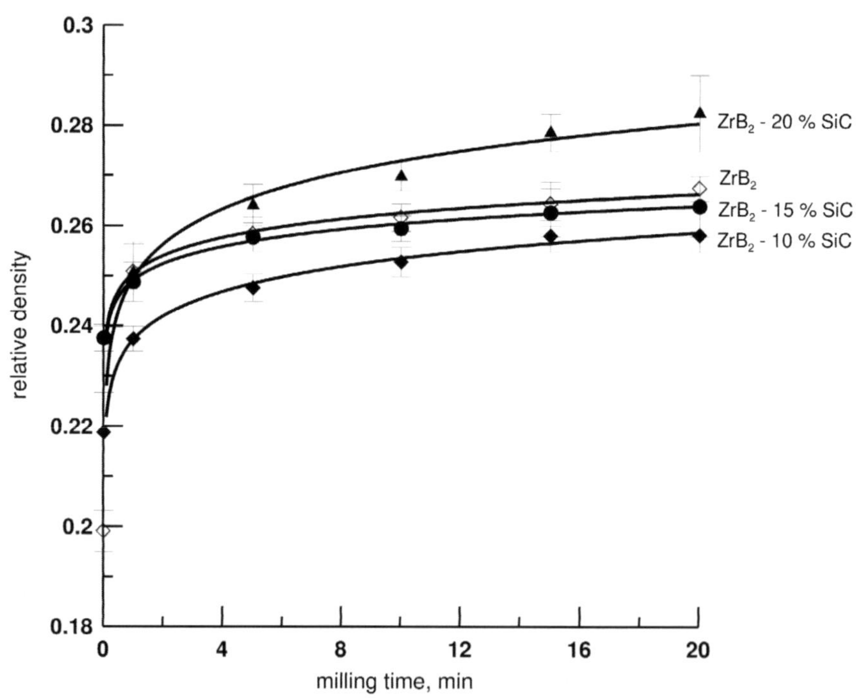

Figure 1. The dependencies of raw density of ZrB_2–SiC powders after mechanical treatment in planetary mill

After hot pressing of ceramics, the phase content did not change, and the increase of treatment time in the planetary mill before sintering have no effect on CDD of sintered materials (Figure 3). This means that all defects are annealed during hot pressing process. In Figure 4, the densities of hot pressed samples are shown. Obviously, addition of SiC leads to essentially increased sample density: its value goes up to 99% of a theoretical one for a powder with 20% of SiC, as compared to ZrB_2 going not higher than 76%. This agrees well with data of [3].

The dependencies of relative density change in sintered materials are well described by a simple function like $Y = aX^n$, where parameter n characterizes the speed of density change. This can be easily determined by re-plotting these dependences in log-log coordinates. Then, the inclination of a straight line will be equal to this parameter. Thus determined parameter "**n**" is shown in Figure 5, and as one can see, the addition of SiC to the mixture leads to up to four-fold decrease of its value. Hence, beside the sample increased density after hot pressure, the material densifies at initial stages of the sintering process.

Original 10 min 20 min

Figure 2. Change of powder morphology during milling

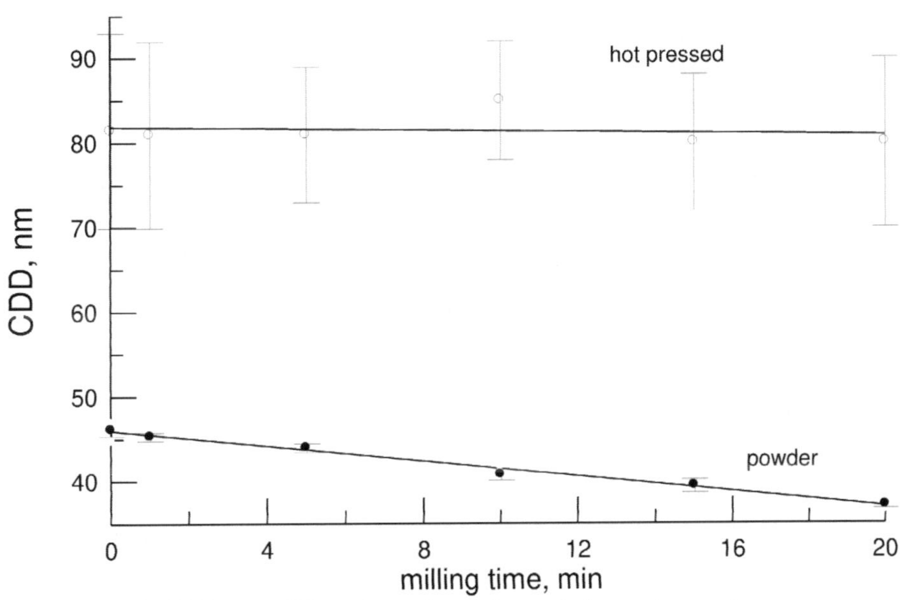

Figure 3. CDD vs. milling time of powder and sintered ceramics.

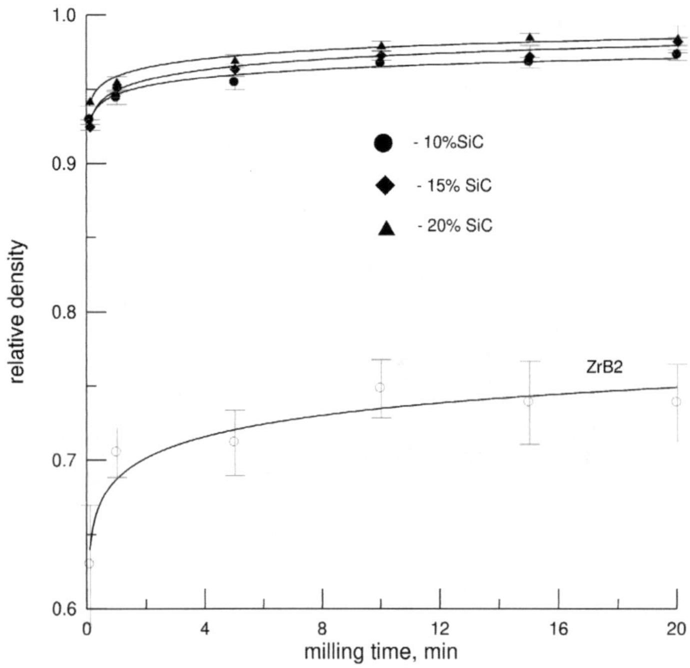

Figure 4. Relative density ρ/ρ_t of hot pressured ceramics vs. milling time of powder.

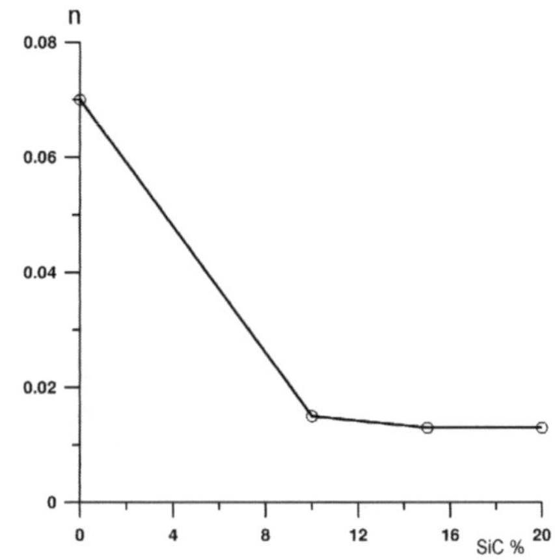

Figure 5. The dependence of parabolic index n of sintering process on SiC content in powder

4. Summary

It was demonstrated that material densification after mechanical treatment is finished at initial stages of sintering process. Addition of SiC leads to essentially increased sample density up to 99% of theoretical value for powder with 20 % SiC, as compare to ZrB_2 with the density not higher than 76%. It was shown that all defects that were accumulated during mechanical treatment are annealed during hot pressing process, and there are no changes of CDD values in sintered ceramics.

Acknowledgement

This work was financially supported by the Ministry of Education and Science of the Russian Federation in the framework of Agreement No. 14.607.21.0056 (RFMEFI60714X0056).

References

[1] Patel M, V Singh, J Reddy, V Bhanu and Jayaram V 2013 *Scripta Materialia* **69** 370–373
[2] Mashhadi M, Khaksari H and Safib S 2015 *Journal of Material Research and Technology* **4** 416-422
[3] Sreckovic T 2006 *Advances in Science and Technology* **45** 619-628
[4] Akin I, Hotta M, Sahin F, Yucel O, Goller G and Goto T 2009 *Journal of the European Ceramic Society* **29** 2379–85
[5] Torayda H, Yoshimura M and Somiya S 1984 Journal of the American Ceramic Society **67** 119-121

Structure and thermal behavior of zirconium tungstate under heating

E S Dedova[1,2,3] E S Shutilova[1] R Geber[4] L A Gomze[4] and S N Kulkov[1,2,3]
[1]Institute of Strength Physics and Materials Science SB RAS, Tomsk, Russia
[2]National Research Tomsk Polytechnic University, Tomsk, Russia
[3]National Research Tomsk State University, Tomsk, Russia
[4]University of Miskolc, Miskolc, Hungary

E-mail: lsdedova@yandex.ru

Abstract. The morphology and properties of powders $ZrW_2O_7(OH)_2{\cdot}2H_2O$ and ZrW_2O_8, obtained under the conditions of hydrothermal synthesis was studied. Using the high-temperature X-ray analysis, the mechanism of formation of zirconium tungstate was established. The influence of temperature on the structure and properties of materials was studied using shadow-casting method.

1. Introduction

Materials with negative thermal expansion have received attention of researchers in recent decades. Scientific interest determined the establishment of the causes and explanation of the unique thermal behavior of this group of materials. Materials contracting upon heating can solve the technical problem towards the incompatibility of thermal expansion of the constructional design elements. The combination of materials with positive and negative values of thermal expansion in the required ratio allows obtaining materials with low or zero thermal expansion (*so-called invar effect*). The field of application of composite materials is wide and includes such areas as the production of high-precision optical mirrors, thermal protection of descent modules. Additive technology enables the production of ceramic components of complex shape.

Zirconium tungstate is a part of oxide complex system AM_2O_8 (A = Zr, Hf; M = W, Mo). A distinctive characteristic of zirconium tungstate is isotropic compression in a wide temperature range from 0.3 K to 1050 K ($\alpha = -8.6{\cdot}10^{-6}$ K^{-1}) [1]. The thermal behavior of the material is due to the structure. The zirconium tungstate structure is known to consist of ZrO_6 octahedron and WO_4 tetrahedron rigidly connected by a common oxygen atom. With increasing temperature, the oxygen atom in Me - O - Me (Me = Zr, W) bonds starts to oscillate in the transverse direction, which causes the rotation of structural elements, leading to compression of materials [1, 2].

The structure of the zirconium tungstate is strongly affected by the production method as well as by external factors. Promising methods for producing nanosized zirconium tungstate powder are chemical methods, including hydrothermal synthesis. In this case, ZrW_2O_8 powder may be prepared by the controlled decomposition of the $ZrW_2O_7(OH)_2{\cdot}2H_2O$ precursor synthesized at relatively low temperatures. The structure of $ZrW_2O_7(OH)_2{\cdot}2H_2O$ plays an important role in the formation of zirconium tungstate [3, 4] and is very sensitive to preparation conditions. Even minor variation in the production process of the powder has effect on the crystallization process of ZrW_2O_8.

Thus, the aims of the work are to study the structure, phase composition and morphology of the precursor and zirconium tungstate powders, to investigate the effect of temperature on the structure and properties of materials, to determine the mechanism of ZrW_2O_8 formation.

2. Experimental procedure

The powder of ZrW_2O_8 was obtained by the thermal decomposition of $ZrW_2O_7(OH)_2{\cdot}2H_2O$ precursor synthesized by the hydrothermal method. The detailed description of the method is given in paper [5].

The specific surface of the powders synthesized was measured by the low-temperature nitrogen adsorption method (BET) [6]. Raw density (ρ_n) was determined by equation: $\rho_n = (m_2-m_1)/V$, where m_1 is the weight of container, m_2 is the weight of container with powder, V is container volume.

The microstructure of ZrW_2O_8 was observed by using Hitachi TM-1000 scanning electron microscope, JEM-2100 transmission electron microscope and MicroVis heating microscope. For this experiment, the powders were compressed into cylindrical samples, which were placed in the installation and heated from room temperature to 1678 K. In microscope, the form of a shadow from samples was fixed, which was used to determine the change of size and shape of simples, wetting angle, the contact angle between the sample and the alumina substrate, on which materials were installed during heating. Scanning was carried out in range of 298–1678 K; the step was 10 K every 10 seconds.

The phase composition and parameters of crystal structure materials was studied using Bruker D8 diffractometer (40 kV, CuKα). Shooting was conducted with the use of standard software and realized in the range of $10 < 2\theta < 70$ with a step size of 0.02°, acquisition interval 3 c. Temperature ranged from 298 to 1023 K with a step size of 100 K.

3. Results and discussion

Scanning and transmission electron microscopy (SEM and TEM) analysis has shown that $ZrW_2O_7(OH)_2{\cdot}2H_2O$ powder was represented by agglomerates of whisker-like particles and single whisker-like particles (Fig. 1a). The average size in transverse dimension of the particles forming the agglomerates was equal to 50 nm, the average length of single particles was 200 nm. Such shape of the particle was due to the influence of initial components, especially the kind of acid. It is known that the whisker shape of $ZrW_2O_7(OH)_2{\cdot}2H_2O$ particles was determined by the use of hydrochloric acid [3, 7].

The zirconium tungstate powder consisted of whisker-like particles having a block structure. The average block size was up to 50 nm (Fig. 1b). The average transverse size of particles was up to 700 nm; the size in longitudinal direction varied between 500 nm and 5 microns. The particle size distribution was unimodal [8].

Evidently, Figure 1 shows that ZrW_2O_8 crystals inherited the shape and size of $ZrW_2O_7(OH)_2{\cdot}2H_2O$ crystals, except for typical nanosized block structures formed in ZrW_2O_8 particles.

Raw density, characterizing the weight of a unit of freely poured powder, for the two powders was almost similar and was equal to 0.3 g/cm^3. The specific surface area, which characterized the dispersion of powder material, for the two powders was significantly different and was equal to 5.58 m^2/g for ZrW_2O_8 and 19.95 m^2/g for $ZrW_2O_7(OH)_2{\cdot}2H_2O$. Probably, the variation in values of specific surface area was determined by the change in the shape of particles, namely the disappearance of agglomerates in precursor powder and formation of blocks in ZrW_2O_8 particles.

Investigation of the effect of temperature on the structure and properties of powders was carried out using the heating microscope. The obtained images helped to determine contact angle, hot stage of samples and estimate the temperatures of the beginning of sintering, softening, melting and spheroidization processes (Figure 2). The obtained values for $ZrW_2O_7(OH)_2{\cdot}2H_2O$ and ZrW_2O_8 are presented in Table 2.

Samples prepared from precursor powder were sintered at lower temperature (958 K) as compared to ZrW_2O_8 (1240 K). Moreover, this sample sustained higher temperatures and melted at 1431 K in contrast to 1360 K for ZrW_2O_8.

a) b)

Figure 1. The SEM and TEM images of a) $ZrW_2O_7(OH)_2{\cdot}2H_2O$ and b) ZrW_2O_8 powders

Table 1. The characteristics of $ZrW_2O_7(OH)_2{\cdot}2H_2O$ and ZrW_2O_8 powders

Samples	Bulk density, g/cm^3	Specific surface area, m^2/g	Average particle size, μm	
$ZrW_2O_7(OH_1,Cl)_2{\cdot}2H_2O$	0.31	5.58	longitudinal	0.5 - 7
			transverse	0.05 – 0.2
ZrW_2O_8	0.31	19.95	longitudinal	0.5 - 5
			transverse	0.03 – 0.7

While the temperature increased up to 473 K, the monotonic decrease of sample's height $ZrW_2O_7(OH)_2{\cdot}2H_2O$ was noted, which continued up to 1408 K; then, there was a softening of the material. The height of the sample decreased by 37% in relation to its original size. The height of ZrW_2O_8 sample slightly changed with increasing temperature.

The dependence of the change in the wetting angle (θ) of the samples with substrate is shown in Figure 3. According to the plot, the wetting angle was remained almost unchanged up to 900 K. Further increase of the temperature up to 1300 ± 23 K led to the decrease and, subsequently, to the increase of θ values, associated with the deformation of samples upon heating. Obviously, Figure 2 demonstrated the change in the shape of the sample under the influence of temperature; the contact area between the sample and the substrate was decreased, which led to the fluctuation of samples. The subsequent increase of the temperature resulted in a sharp decrease of wetting angles associated with melting of the materials.

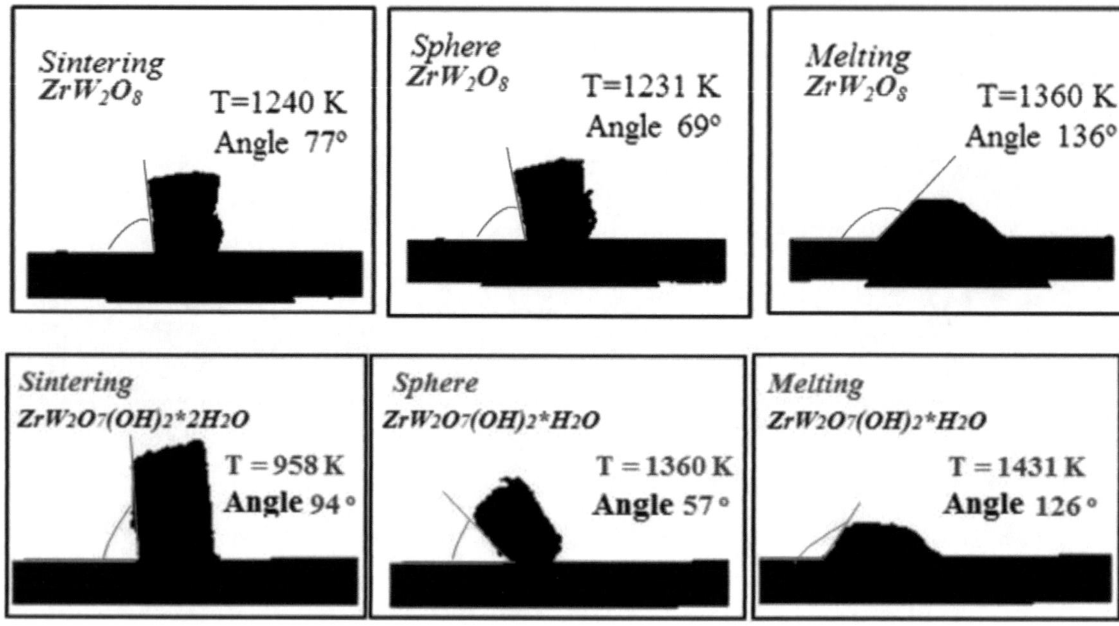

Figure 2. Heating microscopy of the evolution of $ZrW_2O_7(OH)_2 \cdot 2H_2O$ and ZrW_2O_8 samples

Table 2. Temperature data for $ZrW_2O_7(OH)_2 \cdot 2H_2O$ and ZrW_2O_8 samples

Samples	Sintering	Softening	Sphere	½ sphere	Melting
$ZrW_2O_7(OH_1,Cl)_2 \cdot 2H_2O$	958 K	1408 K	1360 K	1405 K	1431 K
ZrW_2O_8	1240 K	1285 K	1231 K	1350 K	1360 K

High-temperature *in situ* X-ray analysis showed that only X-ray peaks of $ZrW_2O_7(OH)_2 \cdot 2H_2O$ were recorded at room temperature [5]. The crystal structure of $ZrW_2O_7(OH)_2 \cdot 2H_2O$ transformed into X-ray amorphous phase above 472 K. The crystal structure corresponding to cubic phase of ZrW_2O_8 was formed up to 873 K. The dependence of the total intensity of all crystalline phases reflexes related to background intensity ($\Sigma I/I_b$) on the temperature in logarithmic coordinates is shown in Figure 4. The resulting dependence was approximated by linear functions; the intersection point of lines corresponding to temperature was equal to 625 K. The comparison of high-temperature X-ray results with thermal analysis carried out earlier [9] has shown that zirconium tungstate was formed by X-Ray amorphous phase at 625 ± 25 K, Fig. 4.

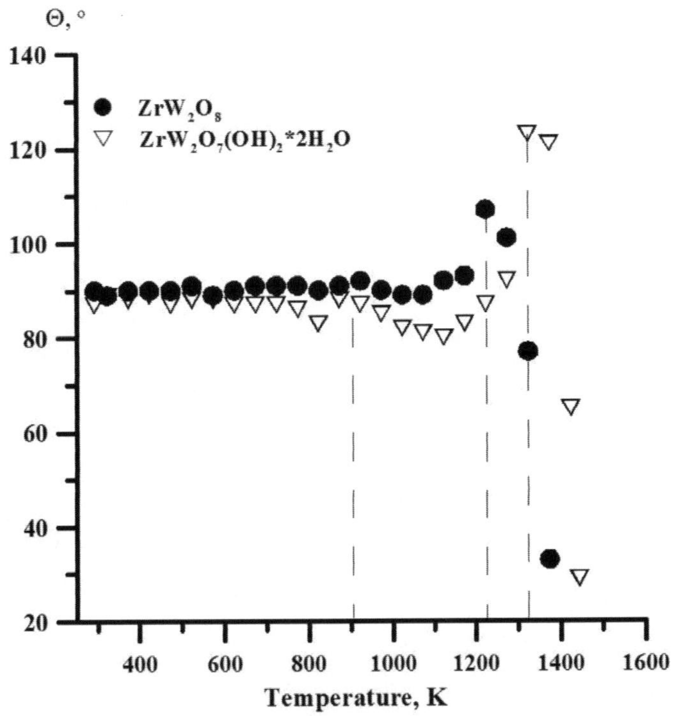

Figure 3. Wetting angle as function of temperature.

Results of high-temperature *in situ* X-ray study of ZrW_2O_8 powder showed solely the peaks of cubic zirconium tungstate in the pattern at room temperature. The heating of the powder above 473 K led to the disappearance of some peaks characteristic to α-ZrW_2O_8. According to [1, 2], the disappearance of some reflexes was due to "order - disorder" phase transformation from low- to high-temperature modifications of zirconium tungstate. Above 823 K, WO_3 and ZrO_2 peaks were recorded in the pattern; their intensity increased with increasing temperature. Reflexes of ZrW_2O_8 were not observed above 1098 K.

The total intensity (ΣI) of ZrW_2O_8 peaks as the function of temperature is shown in Figure 5. As one can see, values ΣI changed with temperature: ΣI decreased from 298 to 473 K, varied within the limits of error from 473 to 823 K and increased above 823 K. The change of total intensity at 473 K corresponded to $\alpha \rightarrow \beta$ phase transition, which was accompanied by the disappearance of diffraction maximum. High-temperature cubic phase of zirconium tungsten remained stable up to 823 K. A further increase of temperature led to an increase in the total intensity of X-ray peaks caused by the advent of WO_3 and ZrO_2 peaks. Possibly, at 823 K, ZrW_2O_8 atoms in the lattice started rearranging to form sublattices of ZrO_2 and WO_3, which led to the decomposition of zirconium tungstate to tungsten (VI) oxide and zirconium (IV) oxide above 1098 K.

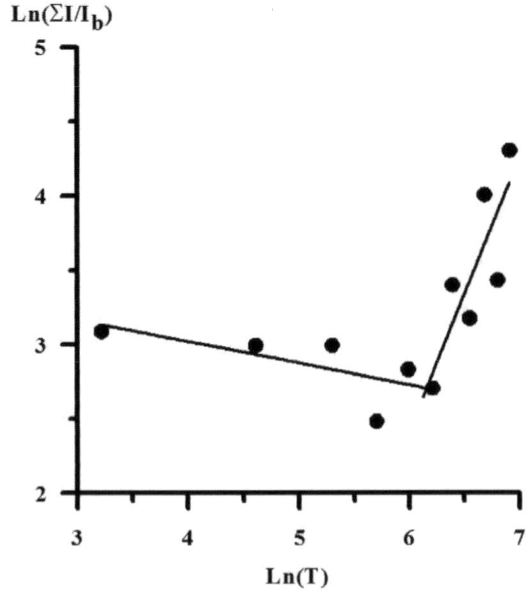

Figure 4. Total intensity of crystal reflections of $ZrW_2O_7(OH)_2 \cdot 2H_2O$ as function of temperature in logarithmic coordinates

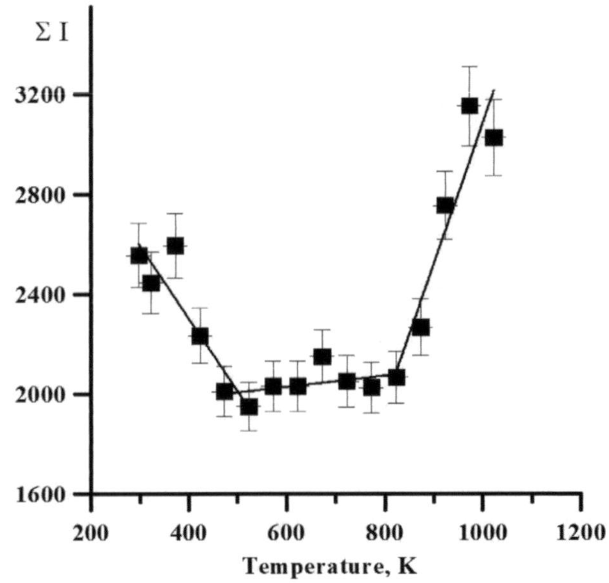

Figure 5. Total intensity of all reflections of ZrW_2O_8 as function of temperature

4. Conclusions

The morphology of $ZrW_2O_7(OH)_2 \cdot 2H_2O$ and ZrW_2O_8 powders was similar and consisted of whisker-like particles, what evidenced the isomorphism of crystals.

The effect of temperature on the structure and properties of powders was investigated. The change of the size and shape of samples, wetting angle, contact angle between the sample and the substrate were determined. It was established that the wetting angle remained almost unchanged up to 900 K. A further increase of the temperature up to 1300 ± 23 K led to the decrease and subsequently the increase of θ values, associated with material spreading on the substrate upon heating.

High-temperature *in situ* X-ray studies have shown that zirconium tungstate was formed through X-Ray amorphous phase at 625 ± 25 K and remained stable from room temperature up to 823 K. Further increase in temperature led to the decomposition of zirconium tungstate caused by the change of the lattice structure by restructuring ZrW_2O_8 atoms to form sublattices of WO_3 and ZrO_2.

Acknowledgements

The research was conducted with partial financial support within the agreement with the Ministry of Education *No. 14.575.21.0040 (RFMEFI57514X0040)*. Transmission electron analysis was obtained using JEM-2100 in the "NANOTECH" Center of ISPMS SB RAS. Study of the morphology (scanning electron microscopy) and thermal behavior (heating microscope) was carried out using analytical equipment of the Miskolc University (Miskolc, Hungary).

References

[1] Mary T A, Evans J S O, Vogt T, Sleight A W 1992 *Science* **272** 90-92
[2] Evans J S O, David W I F, Sleight A W 1999 *Acta Crystallographica* **330** 333–340
[3] Kozy L C, Tahir M N, Lind C, Tremel W 2009 *J Mater Chem* **19** 2760 – 2765
[4] Lind C, Wilkinson A P, Rawb C J, Payzant E A 2001 *J Mater Chem* **11** 3354 - 3359
[5] Gubanov A I, Dedova E S, Plysnin P E and etal 2014 *Thermochimica Acta* **597** 19-26

[6] Greg S and Sing K 1982 *Adsorption Surface Area and Porosity Academic Press*
[7] Badrinarayanan P, Ahmad Md I ,Aknic M, Kessler M R 2011 *Mater Chem. and Phys* **131** 12 – 17
[8] Kulkov S N, Dedova E S, Pedraza F, Erdelyi J 2014 *Epitoanyag - J Sillcate based and Comp Mater* 66(2) 35 - 37
[9] Kulkov S N, Dedova ES, Gubanov A 2013 *I Izv. Vyssh. Ucheb. Zaved. Fizika* **56** 151–156

IOP Conf. Series: Materials Science and Engineering **140** (2016) 012008 doi:10.1088/1757-899X/140/1/012008

Numerical Study of the Features of Ti-Nb Alloy Crystallization during Selective Laser Sintering

A I Dmitriev[1,2,3*] and A Y Nikonov[1,2]
[1]Institute of Strength Physics and Materials Science, SB RAS, Tomsk, Russia
[2]National Research Tomsk State University, Tomsk, Russia.
[3]National Research Tomsk Polytechnic University, Tomsk, Russia.

E-mail: dmitr@ispms.ru

Abstract. The demand for implants with individual shape requires the development of new methods and approaches to their production. The obvious advantages of additive technologies and selective laser sintering are the capabilities to form both the external shape of the product and its internal structure. Recently appeared and attractive from the perspective of biomechanical compatibility are beta alloys of titanium-niobium that have similar mechanical properties to those of cortical bone. This paper studies the processes occurring at different stages of laser sintering using computer simulation on atomic scale. The effect of cooling rate on the resulting crystal structure of Ti-Nb alloy was analysed. Also, the dependence of tensile strength of sintered particles on heating time and cooling rate was studied. It was shown that the main parameter, which determines the adhesive properties of sintered particles, is the contact area obtained during sintering process. The simulation results can both help defining the technological parameters of the process to provide the desired mechanical properties of the resulting products and serve as a necessary basis for calculations on large scale levels in order to study the behaviour of actually used implants.

1. Introduction

Currently, the dominant materials for producing bio-implants are titanium and titanium alloys. Along with the recognized advantages of titanium as the implant material (magnetic inertia, biocompatibility, low density combined with high strength, high corrosion resistance and others), titanium has a relatively high modulus of elasticity. It was found that from attractive perspective of high biomechanical compatibility, beta alloys of titanium-niobium may be considered, which may also have the effect of pseudoelasticity bringing them closer in properties to the properties of cortical bone [1, 2]. Metallurgical methods for producing such alloys are well known but they are complicated, laborious and expensive. This is caused, in particular, by a large difference between the melting points and densities of titanium and niobium. It should be noted that titanium and niobium are significantly different in thermal properties (thermal conductivity, specific heat, coefficient of linear expansion). Manufacturing also requests complex multistep thermomechanical methods in order to generate alloys with homogeneous structure and chemical composition. Furthermore, after such treatment, up to 50% of preform volume is lost.

---

* To whom any correspondence should be addressed.

The additive technologies being developed over recent years, in particular, the methods of layerwise laser synthesis, allow solving the abovementioned problems and accomplish the objectives of personalized medicine [3, 4]. Specific advantages of additive technology and selective laser sintering are the capabilities of forming not only the volume of the product, but also the internal structure. When forming structure, it is possible to simultaneously synthesize a new material (alloy) using the components of initial powder mixture in desired proportions.

Obviously, the physical properties of samples are determined primarily by processing conditions and the properties of the initial powder. In turn, the problem of determining the optimal impact parameters on the initial powder to produce samples with desired mechanical properties requires a deep understanding of mechanisms that are taking place during laser sintering at different scale levels. Picosecond duration and high power of laser beam makes the direct experimental study of the effects of a high-energy impact appreciably difficult. One of the research directions allowing us to identify the main regularities of the formation of Ti-Nb alloy by layering laser synthesis, depending on the parameters and intensive modes of heating, is the use of numerical modelling. Thus, using the numerical method of *ab initio* calculations described in [5], the optimal stoichiometry for obtaining the desired elastic modulus for Ti-Nb alloy system was found. In the present paper, a mathematical model of the process of low modulus β-Ti-Nb alloy system formation produced by laser sintering from titanium and niobium powders was developed. To do so, the method of atomic-scale particles (method of molecular dynamics) was used to model the process of alloy crystallization at different temperature gradients. Alteration of the temperature gradients was determined by the regimes of initial temperature state of the system and different intensity of heat removal determined by thermophysical properties of the substrate. Further, the adhesion between the particles of obtained Ti-Nb alloy determined by the duration of the high-energy laser exposure was estimated.

2. Calculation Results

2.1. Features of Ti-Nb alloy crystallization

To study the features of Ti-Nb alloy crystallization under different temperature gradient conditions on the scale of single atoms, the method of molecular dynamics in the form of LAMMS software [6] was used. Interatomic interaction was described using an interatomic potential calculated using the embedded atom method [7, 8], which earlier had been verified in test problems by calculating elastic and energy characteristics. The simulated system consisted of two spherical crystallites, one of which corresponding to the lattice of titanium, and the second one, to the lattice of niobium. The initial sample structure is shown in Figure 1a.

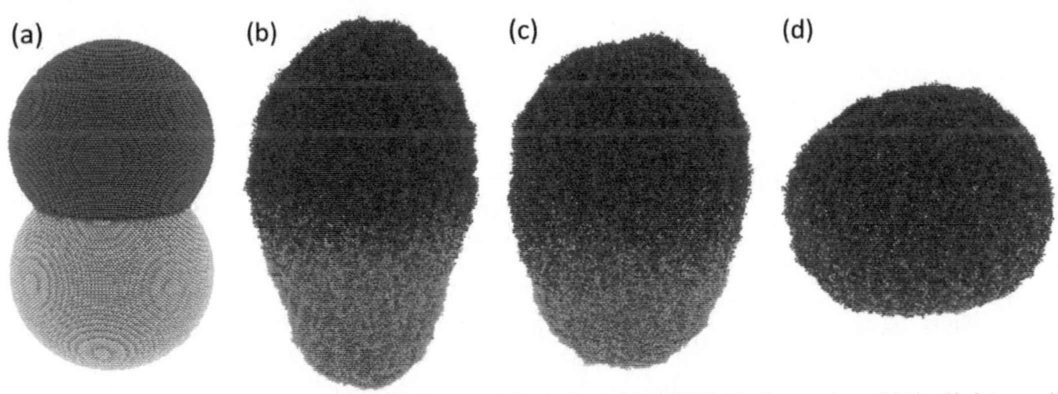

Figure 1. (a) The initial structure of the modeled particles of Ti (dark gray) and Nb (light gray) and structure corresponds to different moments of time of the high-speed heating process up to 6000K: (b) 70 ps, (c) 120 ps and (d) 220 ps.

The diameter of each sphere was 10 nm, which is 5 times less than the characteristic size of the powder particles used for laser synthesis. The total number of atoms exceeds 150000. In order to study the effect of varying duration of a laser beam exposure, the high-speed warm-up of two particles up to temperature of 6000 K, 5000 K and 3000 K in various tasks was realized. Snapshots given in Figures 1b and 1c demonstrate the structure of the modeled system at different moments of time corresponding to the heating process. Because of warm up stage, the uniform melting for both metals within one particle was achieved as shown in Figure 1d.

The future solidification procedure, during which a linear decrease of the system temperature from a high value down to ambient one (300 K) was carried out. The following cooling rates of the modeled ensemble of atoms were studied: 9.5, 5.7 and 3.1 deg./ps. The corresponding time dependencies of the temperature changing for the modeled system during cooling stage are shown in Figure 2. Note that in order to achieve the cooling rate of 3.1 deg./ps, the resulting temperature of the system after heating stage was decreased down to 5000 K. The resulting structure of the sample after cooling was analyzed using the algorithm, which allows determining the local topology of interatomic structure corresponding to bcc, fcc and hcp crystal lattices. The influence of the cooling rate on the fraction of atoms with bcc local topology of Ti-Nb alloy was studied.

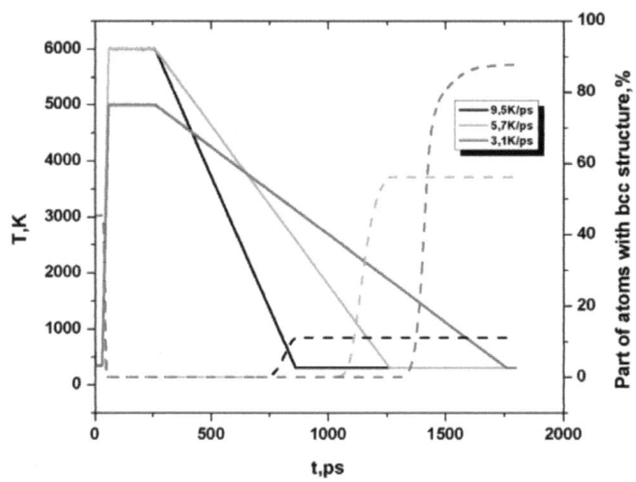

Figure 2. The results of crystallization process for Ti-Nb system. Time dependencies of the system temperature and fraction of fcc atoms for three cooling rates.

The resulting structures for considered variants of cooling rates are shown in Figures 3a-3c. The percentage of bcc atoms for the resultant state depends on the cooling rate and amounts to 11.2%, 56.2% and 87.7%, respectively. Further reduction of the system cooling rate did not result in a noticeable change in the proportion of atoms with bcc lattice in the resulting state. Thus, by trying different cooling modes of the system, we found that the cooling rate of 3.1 deg./ps or less for two-component system is sufficient to ensure the formation of the resulting polycrystalline structure of Ti-Nb beta-alloy. The remaining 12% correspond to the surface atoms, and atoms located near the formed grain boundaries.

These results allow using mechanical and physical characteristics of the formed particles of Ti-Nb alloy as the parameters needed to describe the interaction between particles for models of mesoscopic scale. It is also noteworthy that there is a qualitative conformance of modeling results with obtained experimental data on the influence of preheating of the powder mixture and usage of substrate for producing beta-titanium-niobium alloy by laser sintering.

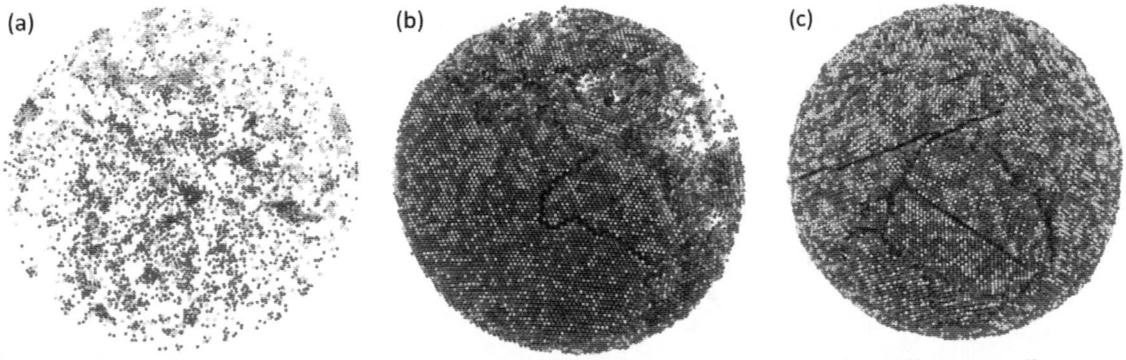

Figure 3. The final structure of the modeled sample after crystallization with different cooling rates: (a) 9.5 deg./ps, (b) 5.7 deg./ps and (c) 3.1 deg./ps.

2.2. Adhesion properties of Ti-Nb particles

Next stage of research was devoted to studying the influence of the heating duration of alloy particle system and its cooling rate on the value of the resulting adhesive force. For this purpose, two hemisphere-shaped particles with an internal structure corresponding to the structure of β-alloy Ti-Nb system obtained by crystallization were simulated. Testing was carried out as follows. Initially, the hemisphere juxtaposed for the distance between their poles of about 1 nm. The configuration of this modelled system is shown in Figure 4a. The crystallographic orientation of the particles was chosen so that in the area of convergence of the particles a grain boundary was formed.

At the next stage the heating of particles in the area of juxtaposed poles was simulated by adding kinetic energy to the surface atoms considering the ratio 1 eV/ps for the whole system. As a result, melting of the hemispheres and their subsequent sintering occurred. Depending on the duration of heating the sintering area was changing. The resulting structure, corresponding to the contact area of 8500 $Å^2$ is shown in Figure 4b. This configuration was obtained by heating the modelled system in the area of converged tops of particles up to ~3000 K within 50 ps, followed by cooling.

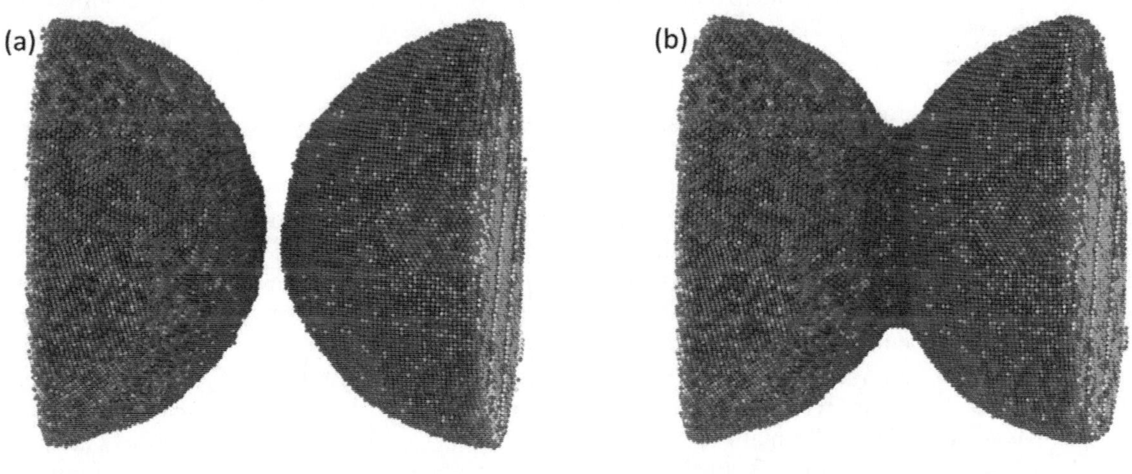

Figure 4. The structure of the modeled setup of Ti-Nb β-alloy at different stage: (a) before heating, (b) after heating.

To assess the adhesive force, the configurations, not only with different values of the contact square, but also with a different structure of resulting contact were created. The modifications in the structure were achieved by changing the cooling rate of the molten area. At low cooling rates, the crystallization of the molten contact area with formation of a grain boundary was observed. At relatively high cooling rates the amorphization of a conjugated area was taking place. The final stage of adhesion force estimation included the simulation of the break of the formed fused region. To do this, the atoms in the base of the hemispheres were assigned constant values of speed oriented in opposite directions along X axis. Figure 5 depicts the dependence of ultimate value of force of resistance to break on the initial area of fusion region. Evidently, there is a clearly pronounced linear dependence between forces and areas for all considered cases. In these particular cases features of structure of the molten area does not have a significant impact on the resulting dependence.

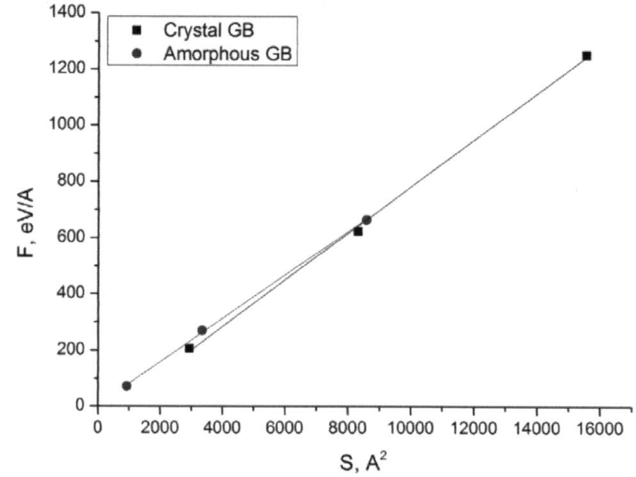

Figure 5. The dependence of adhesion strength on the area and the structure features of the molten area of two particles of Ti-Nb β-alloy.

Thus, the main factor determining the magnitude of the adhesive force is the area of the surface involved in the contact between two particles. This parameter directly depends on the duration of thermal exposure. The cooling rate and the related structure of the molten area has almost no effect on the resulting adhesion properties.

3. Conclusions

The results of computer simulation of melting process, followed by the crystallization of a mixture of particles of titanium and niobium, for different values of the temperature gradient allowed determining the conditions necessary to obtain a system with a maximum content of beta-crystal structure of the alloy Ti-Nb. It was found that the cooling rate of 3.1 deg./ps or less for considered two-component system is sufficient to ensure that the resulting polycrystalline structure will be a beta alloy with maximum content of bcc lattice. The variation of temperature gradients was determined by initial system temperature and various intensity of heat removal defined by thermophysical properties of the substrate. It should be noted that the obtained results qualitatively conform with experimental data on the effect of preheating the powder mixture and the use of substrate during the production of beta-alloy of titanium-niobium system by selective laser sintering [9, 10]. According to the simulation results, it was shown that the main factor determining the value of adhesion forces between the particles is the area of the region involved in the contact between two particles. The rate of cooling and the structure of the obtained contact region has almost no effect on the properties of the resulting adhesion. The obtained result will be used to specify the mechanical and physical characteristics of the formed powder particles of Ti-Nb alloy in the description of the interaction between the particles in the model of a mesoscopic scale.

Acknowledgments
The research work was supported by the Russian Science Foundation, grant No. 15-19-00191.

References

[1] Bai Y, Deng Y, Zheng Y, Li Y, Zhang R, Lv Y, Zhao Q, Wei S 2016 *Mat. Sci. Eng.* C **59** 565

[2] Andrade D P, Vasconcellos L M R, Carvalho I C S, Santos E L S, Prado R F, Santos D R, Cairo C A A, Carvalho Y R 2015 *Mat. Sci. Eng.* C **56** 538-544

[3] Fischer M, Joguet D, Robin G, Peltier L and Laheurte P 2016 Mat. Sci. Eng. C (doi://dx.doi.org/10.1016/j.msec.2016.02.033)

[4] Silva D N, Oliveira M G, Meurer E, Meurer M I, Silva J V L, Santa-Bárbara A 2008 *J. Craniomaxillofac Surg* **36** 443-449

[5] Nikonov A Yu, Zharmukhambetova A M, Skripnyak N V, Ponomareva A V, Abrikosov I A, Barannikova S A, and Dmitriev A I 2015 *AIP Conf. Proc.* **1683** 020165

[6] Plimpton S J 1995 *J. Comp. Phys.* **117** 1-19

[7] Daw M S and Baskes M I 1984 *Phys. Rev.* B **29** 6443

[8] Jones C and Farkas D 1996 *Comp. Mat. Sci.* **6** 231-239

[9] Khimich M A, Parilov E A, Kovalevskaya Zh G and Sharkeev Yu P 2015 *IOP Conf. Ser.: Mater. Sci. Eng.* **93** 012042

[10] Saprykina N A, Saprykin A A, Borovikov I F, Sharkeev Y P 2015 *IOP Conf. Ser.: Mater. Sci. Eng.* **91** 012031

Composite materials based on high-modulus compounds for additive technology

M Grigoriev[1,2], N Kotelnikov[3], S Buyakova[1,2,3], S Kulkov[1,2,3]
[1]Institute of Strength Physics and Materials Science SB RAS, 2/4, pr. Akademicheskii, Tomsk, 634055, Russia
[2] Tomsk Polytechnic University, Tomsk, 30, Lenin Avenue, Tomsk, 634050, Russia
[3] Tomsk State University, Tomsk. 36, Lenin Avenue, Tomsk, 634050, Russia

E-mail: grv@ispms.ru

Abstract. The effect of adding nanocrystalline ZrO_2 and submicron TiC to ultrafine Al_2O_3 on mechanical properties and the microstructure of the composites developed by hot pressing was investigated. It was shown that by means of hot pressing in argon atmosphere at the sintering temperature of 1500 °C one can obtain the composites of Al_2O_3-ZrO_2-TiC with a fine structure and minimal porosity. It was shown that in the material a multi-scale hierarchical structure is formed, which possesses high physical and mechanical properties: the hardness and fracture toughness was 22 GPa and 5.2 MPa*m$^{1/2}$, respectively. It has been shown that mechanical properties of the composite are better than those of commercial composites based on aluminum oxide (Al_2O_3, ZTA, Al_2O_3-TiC) and are comparable to those of silicon nitride.

1. Introduction

Additive technologies are becoming more widely used for various applications. In particular, for example, layerwise printing methods in the aviation and aerospace industry enabled the creation of more complex parts, e.g. turbine blades, turbines etc.

There are various 3D printers that allow printing materials from plastic, metals, ceramic, which is important for the aerospace industry, especially for making very hard and heat resistant engine parts working at high temperatures [1, 2]. For such applications, there are composite materials based on high-modulus compounds like MeO-MeC, where MeO is matrix and MeC are strengthening particles, which increase the heat resistance of the material [3, 4].

Alumina-TiC composites are widely used in industry due to their machinability at higher speeds than cemented carbides and their superior hardness, toughness and strength as compared to alumina [5]. They are commonly known as *black ceramics*, consisting for 70% from alumina and for 30% of TiC, having high hardness of about 22 GPa, but insufficient toughness, not more than 4 MPa*m$^{1/2}$ [6]. Alumina-zirconia composites are also used in industry, where zirconia toughens the alumina matrix by stress induced by tetragonal to monoclinic martensitic phase transformation. The hardness and fracture toughness of Al_2O_3-ZrO_2 composites is only 18 GPa and 5 MPa*m$^{1/2}$, respectively [7]. It is well known that addition of either oxide or non-oxide additives improve mechanical properties of alumina. In this context, addition of both oxide and non-oxide additives may be an attractive option, as it may impart the beneficial effects of both additives to the resulting composites. It becomes even more attractive, if the oxides and non-oxides are nano sized as the majority of the nano particles may remain at grain boundaries and interact with cracks resulting in interesting features not observed in conventional composites.

Thus, the main goal of this paper is to study the structure and properties of ceramic composite materials based on alumina and additives therein of zirconia and titanium carbide in terms of the material applicable for additive technology.

2. Materials and experimental procedure

In these studies, we used commercial Al_2O_3 powders with average particle size of 4.7 μm and with purity of 99.0%, TiC powders with particle size of 5 μm and purity of 98.0% and ZrO_2 powders with nominal particle size of 30 nm and purity of 99.7%.

Figure 1. SEM micrograph of powders: Al_2O_3 (A), ZrO_2 (B), TiC (C)

The mixture of powders was prepared as follows: prepared individual water suspensions of components were mixed with each other with a magnetic stirrer, followed by suspension sonication. The resulting composition was deposited from the solution by flocculation of particles by raising the PH level followed by vacuum drying. The resulting composite mixture is shown in Table 1.

Table 1. The ratio of components in composites

Samples	Content		
	Al_2O_3 [%]	ZrO_2 [%]	TiC [%]
AZT-1	85	10	5
AZT-2	80	10	10
AZT-3	70	10	20
AZT-4	60	10	30

As it was shown earlier [8], the addition of zirconium dioxide of more than 10 wt% leads to reduced hardness of the composites; therefore, the zirconia content was 10 wt%. Ceramic composites were obtained by hot pressing in argon atmosphere at the sintering temperature of 1500 °C, with a pressing under 50 MPa. The holding time was 10 minutes. The heating rate up to hot pressing was 50, 150 and 300 °C/min.

X-ray diffraction data were obtained using an X-ray diffractometer with CuK source; grain-size and elemental analyses of the composites were carried out using LEO EVO 50 (Zeiss, Germany) scanning microscope.

The densities of sintered samples were determined by Archimedes' method with distilled water. For hardness measurements, the sintered samples were polished by diamond paste up to 1 μm grains. Vickers hardness was determined using a 5 kg load with ten indentations for each sample as:

$$Hv = 1.854 \cdot P/d^2 , \quad (1)$$

where P is load [N]; d is the diagonal of indentation [μm].

Fracture toughness was determined as follows:

$$K_{1c} = 0.035 \cdot (H \cdot a^{1/2}) \cdot (E\varphi / H)^{0.4} \cdot (l/a)^{-0.5}, \tag{2}$$

where H is hardness [GPa]; E is Young's modulus [MPa]; a is half-diagonal indentation [μm]; l is the crack length from the corner indentation [μm]; φ is a constant.

3. Results and discussion

Elemental analysis of the fractured surface of Al_2O_3–$10ZrO_2$–20TiC composite has shown that all phases are distributed randomly, but titanium carbide has a higher grain size. Measuring of the average grain size of the individual components has shown that that for alumina is 1.5 microns; for zirconia, 0.8 microns; for titanium carbide, 2.5 microns; thus, the average grain size of individual components in the structure of the composites is not significantly higher than the average particle size of the initial powders.

XRD analysis showed that alumina in composites is in α-modification (corundum), zirconia is cubic and tetragonal modifications of titanium carbide has NaCl-cubic lattice.

Density of the samples was close to the theoretical one, and they had less than one-percent porosity. Evidently, Figure 2 shows that the hardness and fracture toughness of the composites linearly increases with TiC content and is not affected by the heating rate during sintering process.

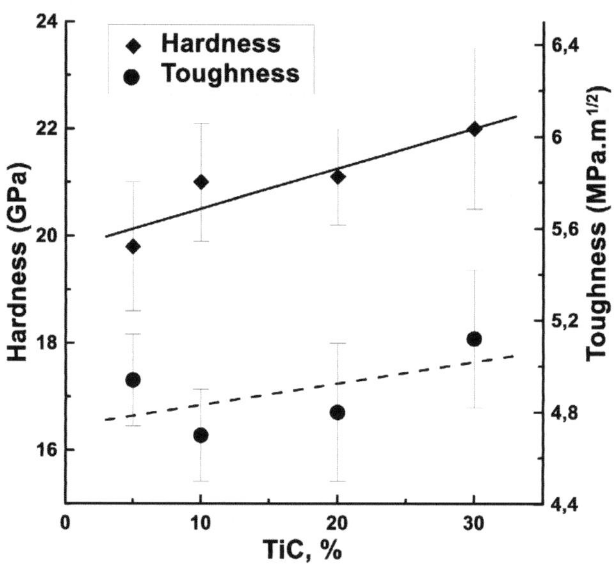

Figure 2. Variation of hardness and fracture toughness vs. titanium carbide content

Such behavior is not in line with data [9, 10], where TiCN and ZrB2 (ZrO2) was added to alumina matrix and had maximum hardness and toughness for TiCN content of 20wt%, and we may expect that the addition of titanium carbide of more than 30% will yield improved hardness and toughness.

The comparison of the strength and toughness of commercially available ceramic composites are shown in Figure 3. Obviously, the mechanical properties of the composite are better than those of commercial composites based on aluminum oxide (Al_2O_3, ZTA, Al_2O_3-TiC) and are comparable to silicon nitride.

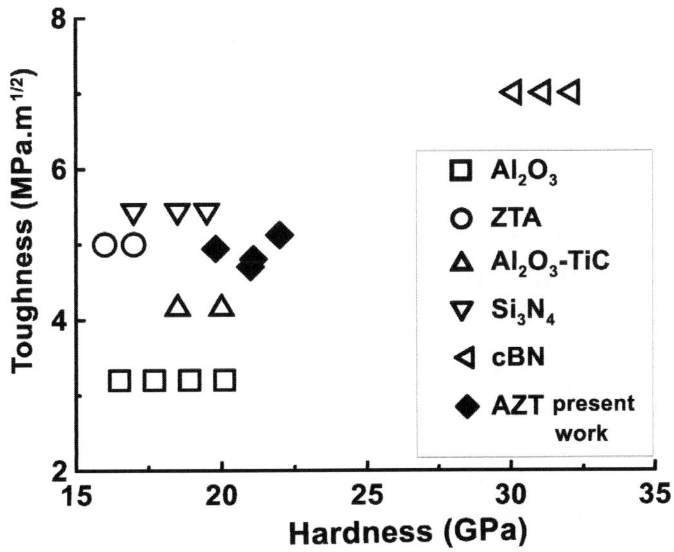

Figure 3. Hardness and toughness of composites compared with different commercial ceramic analogues [11]

4. Conclusions

1. It was shown that hot pressing of Al_2O_3-ZrO_2-TiC composites in argon atmosphere and sintering temperature of 1500 °C yielded very low-porosity and fine structure with uniform distribution of ZrO_2 and TiC phases in alumina matrix without phase changes as compared to initial mixtures.

2. It was demonstrated that hardness and fracture toughness increase linearly with the increase of titanium carbide content in composites, and these mechanical characteristics are very similar to the best values obtained elsewhere.

Acknowledgement
This research was partially supported by RFBR; contract No. 26 16-33-00889\16.

References

[1] Ercin Cura M, Kim SH, Muukkonen T, Varjus S, Vaajoki A, Soderberg O, Suhonen T, Hannula SP 2013 *Ceram Int* **39** 2093–2105
[2] Kumar A S, Durai A R, Sornakumar T 2003 *Int J Refract Met Hard Mater* **21** 109–117
[3] Shaw M C 2005 *New Delhi:Oxford University Press* **2** 307–348
[4] Grigoriev M V, Kulkov S N 2011 Russian Physics Journal **12** 1305–1311
[5] Jianghong G, Hezhuo M, Zhe Z 2001 *J Eur Ceram Soc* **21** 2377–2381
[6] Zhang Y, Wang L, Jiang W, Chen L, Bai G 2006 *J Eur Ceram Soc* **26** 3393–3397
[7] Savchenko N L, Sevostyanova I N, Sablina T Yu, Gomze L, Kulkov S N 2014 *AIP Conf. Proc* **1623** 547-550
[8] Grigoriev M V, Kotelnikov N L, Buyakova S P, Kulkov S N 2015 *AIP Conf. Proc* **1683** 020061
[9] Dibyendu C, Sundararajan G 2013 *J Eur Ceram Soc* **33** 2597–2607
[10] Bin L, Jianxin D, Yousheng L 2009 *Int. Journal of Refractory Metals & Hard Materials* **27** 747–753
[11] Li X S, Low I M 1994 *Key Engineering Materials* **96** 1–18

IOP Conf. Series: Materials Science and Engineering **140** (2016) 012010 doi:10.1088/1757-899X/140/1/012010

Problems of Development and Application of Metal Matrix Composite Powders for Additive Technologies

Elena N Korosteleva[1,2], Gennadii A Pribytkov[2], Maxim G Krinitcyn[1], Anton V Baranovskii[1], Victoria V Korzhova[2]
[1]Tomsk Polytechnic University, 30 Lenin av., 634050, Tomsk, Russia
[2]Institute of Strength Physics and Materials Science of the Siberian Branch of the Russian Academy of Sciences, 2/4, pr. Akademicheskii, 634055, Tomsk, Russia

E-mail:elenak@ispms.tsc.ru

Abstract. The paper considers the problem of structure formation in composites with carbide phase and a metal binder under self-propagating high-temperature synthesis (SHS) of powder mixtures. The relation between metal binder content and their structure and wear resistance of coatings was studied. It has been shown that dispersion of the carbide phase and volume content of metal binder in the composite powders structure could be regulated purposefully for all of studied composites. It was found that the structure of surfaced coating was fully inherited of composite powders. Modification or coarsening of the structure at the expense of recrystallization or coagulation carbide phase during deposition and sputtering does not occur.

1. Introduction
An application of additive technologies to mechanical engineering has great prospects, because it allows significantly reduce of labor input at the manufacturing of complex shape parts while retaining the required quality. However, there are problems of powder feedstock materials for 3D-printing [1-2].

Along with the technological tasks related directly with the process of layerwise shaping articles, there is a problem of control and management of structure formed in this product, which should provide the necessary service properties.

In this regard the studies of interrelations of structural and phase states of composite powders and technological parameters of their preparation with subsequent application in surfacing processes are actual tasks.

The promising materials for additive technologies are metal matrix composite materials with fine carbide phase embedded to the binder [3-6]. In spite of numerous works devoted to carbide-containing powder materials [7, 8], a number of issues remains related to obtaining a predetermined composite structure: carbide particles of desired dispersity and volume content homogeneously distributed in the metal binder.

Results of structure studies of the composite powders "TiC – metal binder", produced by self-propagating high temperature synthesis (SHS) in reactive powder blends of a wide range of inert metallic binder are presented in the paper. Structure and wear resistance of the coatings, overlaid by electron beam facing of the composite powders are presented too.

2. Materials and Experimental Procedures
Powders composed of disperse titanium carbide particles embedded into metal binder were obtained

IOP Conf. Series: Materials Science and Engineering **140** (2016) 012010 doi:10.1088/1757-899X/140/1/012010

by self-propagating high-temperature synthesis (SHS). Reactive powder mixtures of titanium (50-125 μm), carbon black (< 0.1 μm), high-chromium cast iron or high-speed steel (<125μm) were used.

The contents of titanium and carbon in the reaction mixtures were calculated assuming the formation of equiatomic composition titanium carbide. A maximum content of thermally inert metal binder (cast iron, steel or titanium) was limited by the value at which the synthesis in layerwise combustion mode is possible. The mechanical activation of powder mixture has been used additionally for SHS ignition in the compositions with higher content of thermally inert binder.

Porous SHS-compact has been crushed and the necessary for coatings deposition fractions of the composite powder have been extracted by sieving. The synthesized composite powders were used for electron beam surfacing (EBS) and detonation spraying of wear-resistant coatings. Structure studies of SHS-composite powders and electron-beam coatings were carried out by using the equipment of the Centre of collective usage "NANOTECH" of the Institute of strength physics and materials science of Siberian brunch of Russian Academy of Sciences (ISPMS SB RAS). AXIOVERT-200MAT metallographic microscope and "LEO EVO 50" scanning electron microscope were used for of SHS-powder morphology analysis, the microstructure of the SHS compacts and of the EBS coatings. The phase composition was determined by X-ray analysis (Co K_α). For phase identification and calculation of lattice parameters the ASTM X-ray data files and PDW in program were used.

3. Results

3.1. Metal matrix composite powders

Carbides containing composite powders of investigated compositions with different volume content of metallic binder were obtained by crashing of SHS cakes. A representative morphology of the powders can be seen below (figure 1, 2).

c) d)

Figure 1. Morphology of SHS products with different contents of metallic binder (vol. %): high chrome cast iron - a) 22; b) 40; high speed steel - c) 20; d) 40.

At scanning electron microscope patterns (figure 1, 2) and cross sections of powder particles (figure 3) one can see, that SHS-composite powders consist of carbide inclusions surrounded by interlayers of metallic binder. Regardless of the metal binder (cast iron, high speed steel or titanium) an increase of binder content in the reaction blends results in size reducing of the carbide inclusions in the composite structure (figures 1-3).

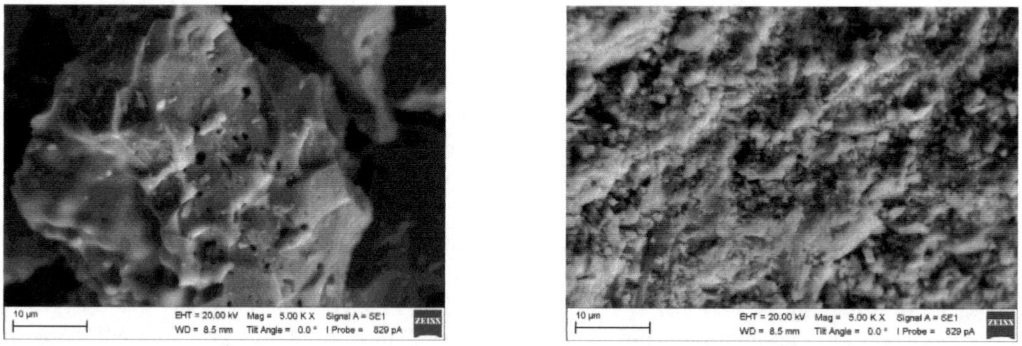

a) b)

Figure 2. Morphology of "TiC – Ti" SHS products in Ti – C reactive mixtures with calculated / real contents of titanium binder (vol. %): a) 30 / 0; b) 60 / 24.6.

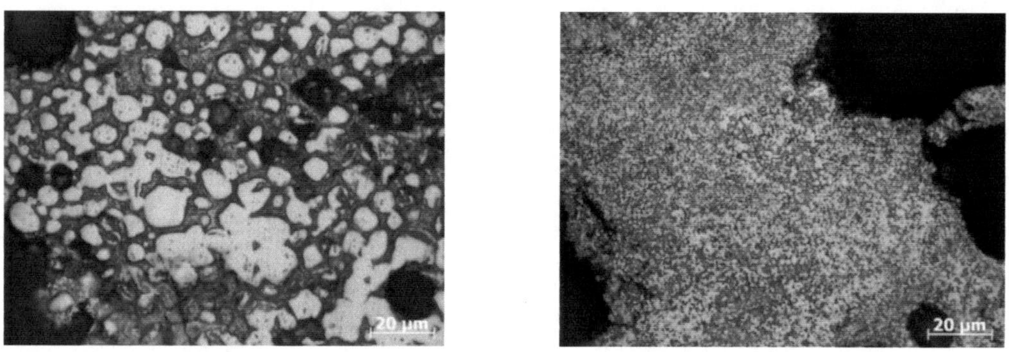

a) b)

Figure 3. The microstructure of SHS composite powder "TiC – high speed steel" with binder content of (vol. %): a) 20; b) 50.

The greater is metal binder content in reactive blends, the less is size of the carbide inclusions. A typical relation "carbide size – metal binder content" presented at figure 4.

An average size of carbide particles in "TiC – high speed steel" composite powders decreases monotonically as the metal binder content in reactive mixes increases. The dependence of the average carbide particles size on metallic binder content in the composite structure can be explained by an influence of thermo-kinetic characteristics of the synthesis on the structure formation of composites [9]. With increasing content of the thermally inert metal binder the maximum combustion temperature decreases [7]. As a result, a temperature interval of the melt existence in a neighborhood of the reaction front becomes more narrow, and the time, during which the structure coarsening occurs by the growth of the carbide particles decreases.

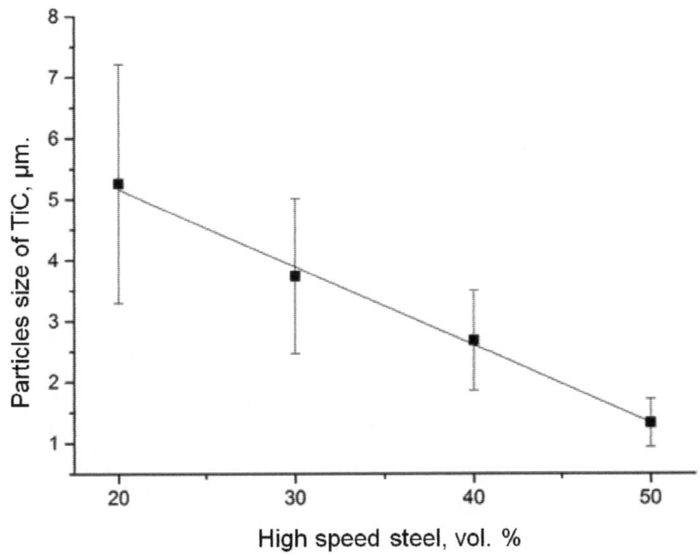

Figure 4. The dependence of the average size of carbide inclusions in SHS composite powder "TiC – high speed steel" on volume content of metallic binder.

A possible deviation from the stoichiometry of the synthesized titanium carbide TiC should be taken into account in carbide-containing metal matrix composites. In particular, titanium binder content in the SHS composite powders "TiC – Ti" does not match with the calculated composition, when it is assumed the formation of equiatomic carbide composition. Owing to binder titanium transition to non-stoichiometric carbide a real binder content is much less then calculated one (Table 1). So in the TiC – 30 vol. % Ti composite the titanium binder is absent at all. Real metal binder content in the rest three composites is twice or more times less then calculated values with the assumption of equiatomic TiC formation (Table 1).

Table 1. The phases content in the "TiC – Ti" SHS products, the lattice parameter *a* of titanium carbide TiC_X and stoichiometry coefficient X, estimated from the lattice parameter.

The calculated content of phases in the SHS powders (with the assumption of equiatomic TiC formation)	The real phases content, %		a, nm	X
	TiCX	Ti		
TiC – 30 vol. % Ti	100	-	0.4320	0.71
TiC – 40 vol. % Ti	95.5	4.5	0.4310	0.58
TiC – 50 vol. % Ti	93.5	6.5	0.4302	< 0.53
TiC – 60 vol. % Ti	75.4	24.6	0.4299	< 0.53

The reason of binder deficit in "TiC – Ti" composites is a formation of non-stoichiometric titanium carbide at the SHS process. According to equilibrium binary diagram at the eutectic temperature titanium carbide with composition close to $TiC_{0.5}$ is in equilibrium with β-Ti base solid solution. So the main portion of titanium from reaction mixture is consumed at the carbide formation.

3.1. EBF coatings overlaid with the composite powders

SHS composite powders containing different metal binder (high chromium cast iron, high speed steel and titanium) were used for coating deposition by electron beam facing (EBF). For obtaining the nonporous coatings the metal binder was added to the composite powders to get an integral 75-80 vol. % binder content in the coatings.

The microstructure of EBF coatings overlaid with "TiC – high speed steel" composite powder is presented on Figure 5. From a comparison of Figure 3 and Figure 5 it follows that dispersion of the carbide phase does not change at surfacing.

This means that the carbide phase recrystallization does not occur when surfacing as a result of dissolving the initial carbide in molten bath and subsequent primary carbide precipitation from the liquid solution under cooling. Thus, the coating structure can be purposefully controlled through the structure of the deposited powder.

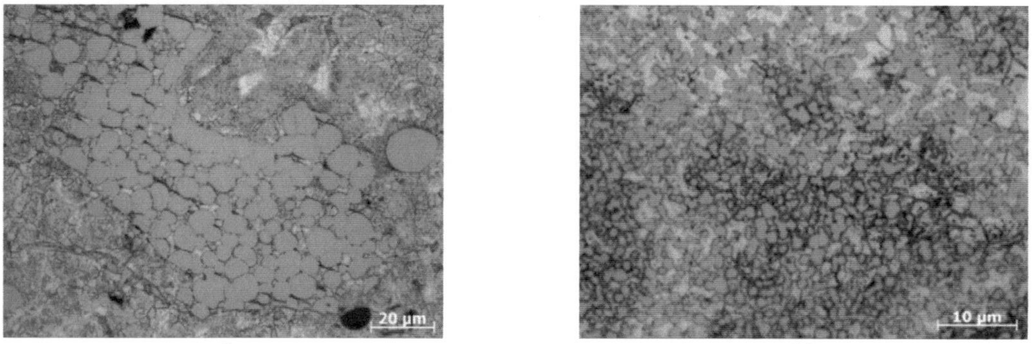

a) b)

Figure 5. Microstructure of EBF coating overlaid with "TiC – high speed steel" composite powders with volume content of steel binder (vol. %): a) 20; b) 50.

The coatings structure has a profound effect on the wear resistance. According to abrasive wear tests of the coatings overlaid with "TiC – high chrome cast iron" composite powders highest wear resistance has the coating overlaid with the powders of high content of metallic binder and low size carbide inclusions (Figure 6). High abrasive wear resistance of the coatings composed of fine carbide inclusions surrounded by narrow binder interlayers can be explained if abrasive wear mechanism is considered. The abrasive wear mechanism involves a deterioration of metal binder interlayers and following tearing away of the carbide particles. If a width of metal binder interlayers is comparable with abrasive grain size, wear rate is high. The narrower is binder interlayer the higher is abrasive resistance of the metal matrix composite. So the coatings with wide metal binder interlayers have a low abrasive wear resistance.

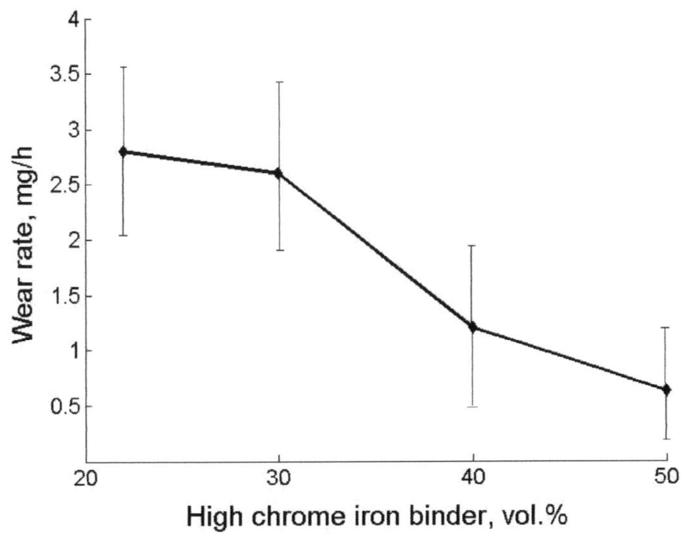

Figure 6. Dependence of electron beam coating wear rate on the binder content in the TiC – high chrome iron composite powders

Summary

1. SHS in powder mixtures of titanium, carbon and thermally inert metals makes possible to obtain composite powders "TiC – metal binder" for deposition of wear-resistant coatings and for using in additive technologies. The elemental composition of the metallic binder can be maintained with good accuracy, because other carbide-forming metal powder components do not participate in the synthesis reaction due to high heat of TiC formation.
2. Dispersion, morphology and volume fraction of the carbide phase in SHS composite powders can be varied within a wide range due to a change of metallic binder content in reaction mixtures and by mechanical activation of the reaction mixtures.
3. A dispersity of the carbide inclusions remains invariable under electron beam surfacing of composite powders. So it is possible to obtain a predetermined structure of the deposited coatings.
4. The abrasive wear resistance of coatings deposited by electron beam facing with "TiC – metallic binder" composite powders depends on the volume content and dispersion of carbide phase, and is 6 ÷ 35 times higher than the wear resistance of VT1-0 titanium or high chromium cast iron.

References

[1] Herderick E 2011 *Proc. Conf. on Materials Science and Technology (October 16-20, 2011, Columbus, Ohio USA)* (Wiley) pp 1413-25
[2] Gong X, Anderson T, Chou K 2014 *Manufacturing Rev.* 1 2 EDP Sciences (2014) 001
[3] Colaco R, Vilar R 2003 *Wear* **255** 643-650
[4] Dogan O N, Hawk J A, Tylczak J H, Wilson R D, Govier R D 1999 *Wear* **229** 758-769
[5] Dogan O N, Hawk J A, Tylczak J H 2001 *Wear* **250** 462-469
[6] Pribytkov G A, Khramogin M N, Durakov V G, and Korzhova V V 2008 *Welding international* **22 7** 465-467
[7] Kalambaeva S S, Korosteleva E N, Pribytkov G A 2014 *IEEE conf. public. Proc.Int.Conf.on Mechanical Engineering, Automation and Control Systems* (Tomsk, Russia)
[8] Krinitcyn M G, Pribytkov G A, Durakov V G 2016 *Key Engineering Materials* **685** 719-723
[9] Rogachev A S, Shkiro V M, Chausskaya I D, Shvetsov M V 1988. *Combustion, Explosion, and Shock Waves*. **24 6** 720-24

IOP Conf. Series: Materials Science and Engineering **140** (2016) 012011 doi:10.1088/1757-899X/140/1/012011

Numerical and experimental study of electron-beam coatings with modifying particles FeB and FeTi

Olga Kryukova, Kseniya Kolesnikova, Nina Gal'chenko

Institute of Strength Physics and Materials Science SB RAS, Tomsk, Russia

E-mail: okruk@ispms.tsc.ru

Abstract. An experimental study of wear-resistant composite coatings based on titanium borides synthesized in the process of electron-beam welding of components thermo-reacting powders are composed of boron-containing mixture. A model of the process of electron beam coating with modifying particles of boron and titanium based on physical-chemical transformations is supposed. The dissolution process is described on the basis of formal kinetic approach. The result of numerical solution is the phase and chemical composition of the coating under nonequilibrium conditions, which is one of the important characteristics of the coating forming during electron beam processing. Qualitative agreement numerical calculations with experimental data was shown.

1. Introduction

Powder metallurgy methods combined with using of concentrated energy fluxes have special capabilities for obtaining materials and compositions with given (improved) properties.

The wear resistant composite coatings on the basis of titanium borides, which are synthesized from thermoreactive powder components of a boron-containing mixture at electron beam surfacing, are interesting for applications today.

It is known that in electron beam surfacing of thermoreactive powders an additional amount of heat is released in the fusion zone due to the exothermal reaction between the mixture components. This allows, with the same electron beam power, the deposition of coatings with a required content of the fine-grained refractory component. The physico-mechanical properties of the coatings can vary in a wide range depending on the proportion and sizes of initial components in the fused mixture. The coating structure formation ends with the convective mixing of solid-liquid melts, which have formed after the complete dissolution of components having different viscosity because of different chemical composition and different content of refractory particles in them [1,2].

The degree of the melt mixing depends on many factors, including the size of initial components. Obviously, changes in the granulometric composition of the fused powder would greatly affect the crystallization rate and completeness of phase transformations at electron beam surfacing, thus defining the microstructure, phase composition and service properties of deposited coating.

The aim of the paper is to study the simultaneous synthesis of boride compounds and formation of composite coatings on their basis under vacuum electron-beam treatment of thermoreactive powders.

2. Materials and experimental procedure

The mixture for the electron beam surfacing of coatings and their further examination was prepared of low-cost thermoreactive ferroalloy powders FeB (grade FB-20) and FeTi (grade FTi65) widely used in steel and alloy metallurgy.

Table 1. Chemical composition of ferroalloys (wt %) used in the study.

Grade	B	Si	Al	C	S	P	Cu	Ti	V
FB-20	20	2	3	0.05	0.01	0.02	0.05	–	–
FTi65	–	1	5	0.4	0.05	0.05	04	65	3

The composite material with a definite proportion of the refractory compound (titanium diboride) and metal binder (iron) was produced by the chemical reaction xFeB + yFeTi → TiB_2 + Fe + Q.
The mixture composition was calculated so that to produce the TiB_2-Fe composite coating with 33 mass % of titanium diboride. The given percentage is provided by the proportion of the used mixture components FeB – FeTi = 1:1.
To study the features of the phase and structure formation in the fused layer, three variants of FeB – FeTi powder mixtures were electron-beam deposited in experiments. The mixtures differed in the granulometric composition of initial components: 1) FeB(50–125 μm)+FeTi(200–315 μm); 2) FeB(200–315 μm)+FeTi(200–315 μm); 3) FeB(200–315 μm) + FeTi(50–200 μm).
The coatings were electron-beam deposited on steel substrates (grade St.3) in 2–4 passes at accelerating voltage 28 kV. The beam diameter was 1.0 mm, scanning length 12 mm, and substrate velocity 2 mm/s. The fused layer thickness was 2–3 mm.
The phase composition was X-ray examined by a diffractometer DRON-4, the chemical composition was studied by a setup KAMEBAX-MIKROBEAM for X-ray spectrum microanalysis on an area of 1 μm. The coating microstructure was examined by a light microscope MIM-9. The microhardness $H\mu$ was measured by a device PMT-3 under the load $P = 50$ g.

3. Experimental results and discussion

3.1. Coatings of thermoactive powder mixture FeB(50–125 μm)+FeTi(200–315 μm) (coating 1)

The comparative analysis has shown that the given coating has the most heterogeneous structure across the thickness. Fragments of the structure are depicted in Figure 1. The structure is typically observed after the crystallization of melts of nonuniform concentration. The uneven microhardness distribution across the coating thickness clearly demonstrates its heterogeneity (Figure 4a).

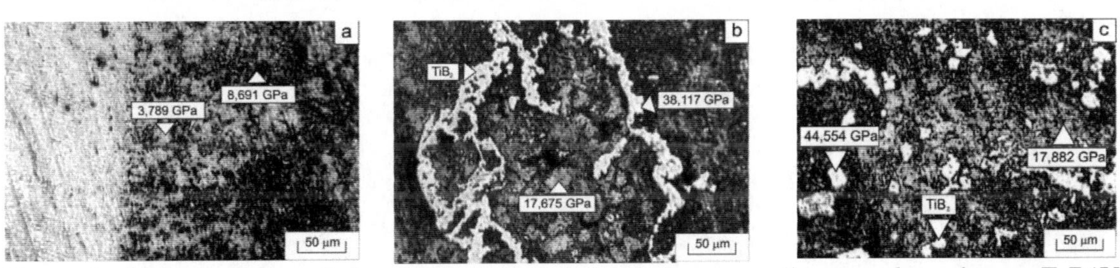

Figure 1. Microstructure of the coating produced of thermoreactive powder mixture FeB(50–125 μm) + FeTi(200–315 μm): *a* — interface with the substrate; *b* — middle of the deposited layer; *c* — subsurface coating area.

The performed investigation suggests the following mechanism of structure formation at surfacing. First, FeTi particles of size 200–315 μm melt ($T_m = 1200$°C) and form a ferrotitanium melt. When contacting with a conglomerate of FeB particles ($T_m = 1540$°C), the melt forms a ring structure around the FeB particles due to the formation of low-temperature contact eutectics. The ring structure consists of titanium diboride crystals of hardness $H = 38$ GPa (Figure 1b). The formation of the liquid phase — ferrotitanium melt — favors further refinement of the FeB particles that rapidly dissolve in the FeTi melt and saturate it with boron. The final coating structure is a result of the solidification of the two heterogeneous melts.

Since the electron beam surfacing is carried out in many passes, the deposition of next layers increases the substrate temperature, lifetime of the liquid metal bath formed on the surface of the underlying deposited layer and reduces the melt crystallization rate. After the crystallization the matrix structure of the composite coating is an alloyed solid solution with highly nonuniform concentration, fine and coarse iron and titanium borides as well as with numerous nonequilibrium boride-based eutectics. This structure leads to the formation of heterogeneous coatings.

Figure 1,c illustrates the microstructure of the FeB(50–125 μm)+FeTi(200–315 μm) coating surface with solitary titanium diboride crystals, which evidently entered the melt due to convective mass transfer from the underlying layer in the middle of the coating. There is a structure on the coating surface which consists of large titanium diboride crystals (7÷22 μm) and heterogeneous metal binder with nonuniform hardness. The binder is chaotically hardened by fine-grained (≤1 μm) compounds of titanium borides of other stoichiometry (Figure 1,c). This is confirmed by the data of X-ray phase analysis showing that, along with pronounced reflections from TiB_2 and α-Fe, phases TiB, Ti_2B and Ti_2B_5 are registered.

3.2. Coatings of thermoreactive powder mixture FeB(200–315 μm)+FeTi(200–315 μm) (coating 2)

The microstructural analysis of coatings produced of a mixture of coarse-grained powders having the same granulometric composition (coating *2*) has also shown that the coating structure is heterogeneous across the thickness (Figure 2).

This is probably due to the fact that, because of the large powder particles and high crystallization rate, the time period when the concentration of alloying elements in the formed melt becomes uniform was too short. The nonuniform melt crystallization resulted in the formation of specific structures at the interface with the substrate and in the middle of the deposited layer. The structures have irregularly shaped regions in the form of light unetched globules whose chemical composition differs across the coating thickness. By the data of X-ray spectrum microanalysis, the above solid solution regions at the interface have the composition of Fe_2B and hardness $H \approx 14$ GPa (Figure 2,a). In the middle of the coating (Figure 2,c) the globules became three-dimensional and darker in color. These regions have the composition of FeB and hardness $H = 16$–18 GPa. The FeB formation is evidently caused by the fact that with distance from the coating – substrate interface and increasing number of passes the melt temperature grows and the degree of its saturation with boron also increases owing to a more complete dissolution of initial FeB powder particles.

Such melt crystallizes with the formation of regions in the coating structure which have the composition of FeB, being separated by fine-grained particles and their conglomerates whose hardness ($H = 32.5$ GPa) corresponds to the TiB_2 phase. One can also see from Figure 2,b,c that the metal binder structure between the unetched regions exhibits heterogeneous phase morphology and nonuniform hardness ($H = 6$–19 GPa). It has numerous regions with dendritic and eutectic structure whose hardness reduces with the eutectic size refinement. According to the X-ray phase analysis, the eutectics can be the Fe_2B-Fe and TiB_2-Fe compounds [3].

When surfacing powders of the given granulometric composition, we observed the highest fluidity of the formed melt as compared to coatings *1* and *3*.

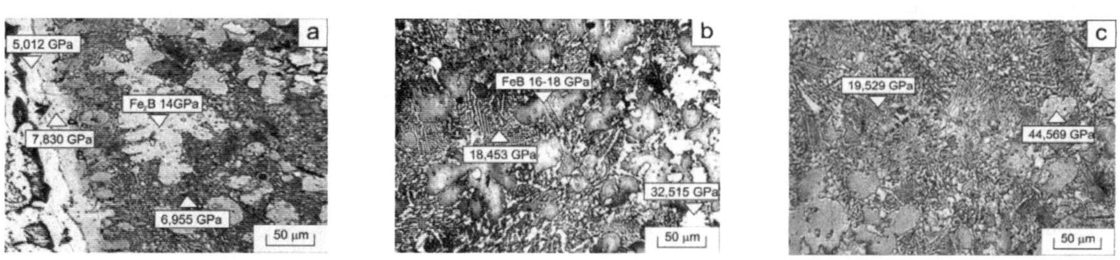

IOP Conf. Series: Materials Science and Engineering **140** (2016) 012011 doi:10.1088/1757-899X/140/1/012011

Figure 2. Microstructure of the coating produced of thermoreactive powder mixture FeB(200–315 μm)+FeTi(200–315 μm): *a*) interface with the substrate; *b*) middle of the deposited layer; *c*) subsurface coating zone.

3.3. Coatings of thermoreactive powder mixture FeB(200–315 μm)+FeTi(50–200 μm) (coating 3)

After coating deposition we had a layered structure, in which every layer mainly contained one phase component across its thickness (Figures. 3, a–c).

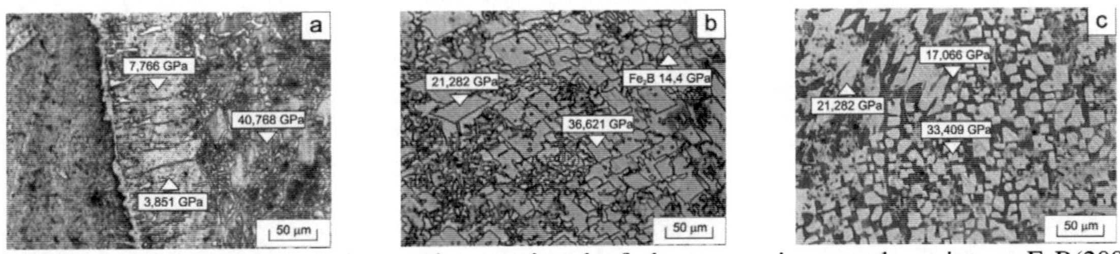

Figure 3. Microstructure of the coating produced of thermoreactive powder mixture FeB(200–315 μm)+FeTi(50–200 μm): *a*) interface with the substrate; *b*) middle of the deposited layer; *c*) subsurface coating region.

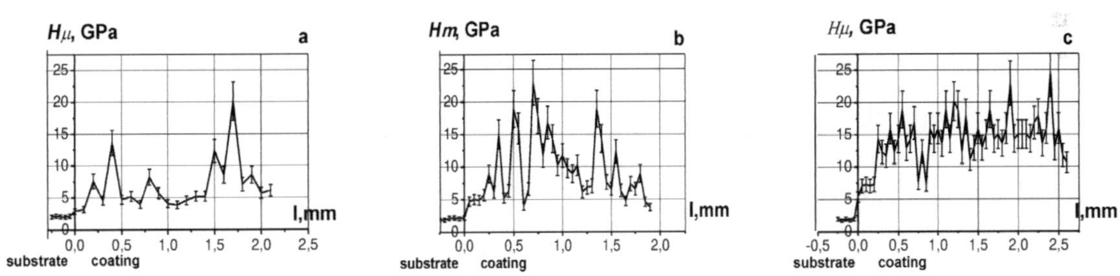

Figure 4. Distribution of microhardness *H*μ in the coating: *a*) FeB(50–125 μm) + FeTi(200–315 μm); *b)* FeB(200–315 μm)+FeTi(200–315 μm); *c*) FeB(200–315 μm)+FeTi(50–200 μm).

In the middle of the coating at 1.3 mm from the interface with the substrate (Figure 3, b) the structure mainly consists of rounded Fe_2B particles of size 8–12 μm and hardness H = 14.4 GPa, rectangular titanium diboride crystals (4–6 μm) and larger elongated titanium boride particles embedded in the Fe_2B-Fe eutectic matrix. The subsurface coating region (Figure 3, c) also has heterogeneous phase composition: the left part of the microscan demonstrates aggregates of titanium borides, the right part — a microstructure region with uniform distribution of 10–17 μm light crystals in the metal binder which are titanium borides and diborides (TiB, TiB_2). The hardness of the particles is in the interval H = 22–34 GPa.

4. Mathematical problem formulation

The mathematical formulation about the phase formation and chemical structure of the coating of the low-carbon steel substrate during electron beam surfacing with modifying particles boron and titanium includes heat equation [4-6]

$$c\rho\frac{\partial T}{\partial t}=\frac{\partial}{\partial x}\left(\lambda\frac{\partial T}{\partial x}\right)+\frac{\partial}{\partial y}\left(\lambda\frac{\partial T}{\partial y}\right)+\Sigma Q_{ch,i}\varphi_i+\sum_{j=B,Ti}Q_{p,j}\Phi_j(T,\eta_{p,j})-\frac{(\varepsilon\sigma T^4-q_e)}{h} \quad (1)$$

and the equation for the particles, which take into account the delivery of the particles in treatment area, turbulent mixing of the particles in the melt under pressure exerted by moving source and dissolution

$$\frac{\partial \eta_{p,j}}{\partial t} = q_{m,j} - \Phi_j(T,\eta_{p,j}) + D_{eff}\left(\frac{\partial^2 \eta_{p,j}}{\partial x^2} + \frac{\partial^2 \eta_{p,j}}{\partial y^2}\right), \quad j = B, Ti \tag{2}$$

where D_{eff} is average mixing ratio of the particles in the melt $D_{eff} = D_T$, $T \geq T_{ph}$; $D_{eff} = D_S$, $T < T_{ph}$; $D_S << D_T$.

The external source moves along the plate surface at the velocity V, and the energy therein is distributed according to the law

$$q_e = \begin{cases} 0 & , y > h_0; \\ q_0 \exp\left(-(x^2 - Vt)/a_t^2\right) & , y \leq h_0, \end{cases} \tag{3}$$

where q_0 is the maximum density of the heat flux; a_t is the effective radius of the source; the parameter h_0 is proportional to the scanning width.

The particle flux density is distributed according to a Gaussian

$$q_{m,j} = q_{m0,j} \exp\left[-\left((x - x_a - Vt)^2 + y^2\right)/a_p^2\right] \tag{4}$$

where q_0 is the maximum density of the heat flux; a_t is the effective radius of the source; the parameter h_0 is proportional to the scanning width.

The rate of dissolution obeys the law $\Phi_j(T,\eta_p) = \varphi_{p,j}(\eta_{p,j}) k_{Sol} \exp\left(-E_{Sol,j}/RT\right)$, where value of activation energy E_{Sol} is determined by features of the dissolution process for particles of a given type. . View of kinetic function depends on the processes that determine the dissolution rate at the micro level. Constant k_{Sol} is determined by the characteristics of the local hydrodynamic flow in the vicinity of the particle trapped in the melt [7].

n the volume of the material the chemical reactions proceed with the total heat release (taking in the heat equation) $Q_{ch} = \rho \sum_{k=1}^{n} h_k \cdot (dC_k/dt)$ where h_k – partial enthalpy (formation) k-th substance, ρ – the material density of the base, C_k – mass concentrations of individual substances and compounds in the mixture, n – number of substances involved in the reactions.

Mass concentrations are related to the molar concentration ratios:

$$C_k = \frac{\rho_k}{\rho}, \text{ since } y_k = \frac{\rho_k}{M_k}, \text{ then } C_k = \frac{y_k M_k}{\sum_{i=1}^{n} y_i M_i + y_{Ti} M_{Ti} + y_B M_B}.$$

Molar concentration we find from solutions of the kinetic problem.

In accordance with the phase diagram of the Fe-B-Ti system we can write a system of chemical reactions:

1). $Fe + B \rightarrow FeB$	4). $Ti + B \rightarrow TiB$
2). $Fe + Ti \rightarrow FeTi$	5). $2Fe + B \rightarrow Fe_2B$
3). $Ti + 2B \rightarrow TiB_2$	6). $2Fe + B \rightarrow TiFe_2$

We introduce the notations for the molar concentrations of the reactants and reaction products involved in the conversion of $y_1 = [\mathrm{Fe}]$; $y_2 = [\mathrm{B}]$; $y_3 = [\mathrm{Ti}]$; $y_4 = [\mathrm{FeB}]$; $y_5 = [\mathrm{FeTi}]$; $y_6 = [\mathrm{TiB}_2]$; $y_7 = [\mathrm{TiB}]$; $y_8 = [\mathrm{Fe_2B}]$; $y_9 = [\mathrm{TiFe}_2]$ mol/m³. Then, for each substance we can concretize equation. For example, for the titanium and boron we have

$$\frac{dy_2}{dt} = \frac{\rho_B}{M_B}\Phi_B - (\varphi_1 + 2\varphi_3 + \varphi_4 + \varphi_5 + \varphi_6 + 4\varphi_7), \qquad \frac{dy_3}{dt} = \frac{\rho_{Ti}}{M_{Ti}}\Phi_{Ti} - (\varphi_2 + \varphi_3 + \varphi_4 + 3\varphi_7).$$

The rate of each reaction depends on the concentration in accordance with the law of mass action, and the rate depends on the temperature in accordance with the Arrhenius law: $\varphi_i = k_{0i}\exp(-E_{ai}/RT)y^{m_i}y^{n_i}$, where i –number of reaction. For many of the written out reactions the experimental data about the formal kinetic parameters are missing. The reaction rates constants k_{0i}, activation energy E_{ai} and heats of reactions Q_i is determined using known thermodynamic formulas [8].

Generally, heat capacity in (1) is complex function of temperature and composition. Near the melting temperature the heat capacity increases rapidly, it represents formula

$$(c\rho)_{eff} = c\rho + L_{ph}\rho_{b\_s}(1-\eta_p)\delta(T - T_{ph}) \tag{5}$$

where δ – the Dirac delta function, L_{ph} – latent heat of melting, T_{ph} – melting point of pure "material of basis"; index "b" corresponds to the material of the base; "s" – the solid phase; "p" – particles.

$$c\rho = (c\rho)_S(1 - \eta_{p\_B} - \eta_{p\_Ti}) + (c\rho)_B\eta_{p-B} + (c\rho)_{Ti}\eta_{p-Ti}, \tag{6}$$

$$(c\rho)_S = (c\rho)_{Fe}\xi_{Fe} + (c\rho)_B\xi_B + (c\rho)_{Ti}\xi_{Ti} + \sum_{k=1}^{n}(c\rho)_k\xi_k,$$

ξ_k is the fraction of newly formed chemical compounds $\xi_k = y_k/(\sum_{i=1}^{n} y_i + y_{Fe} + y_B + y_{Ti})$.

Boundary conditions reflect a lack of heat source at an infinite distance from molten pool and on the end of the plate, adopted as the starting point.

At the initial time we have the conditions

$$t = 0: \quad T = T_0, \qquad \eta_{p,j} = 0, \qquad \eta_{Fe} = 1, \qquad y_{Fe} = 0.141. \tag{7}$$

5. Analysis of results

Analysis of the results of the numerical analysis of the problem shows that as the source moves along treated surface quasistationary regime establishes [4, 9]. The power density of the heat source and the flux density of the particles influence on the shape and size of the molten pool and heat affected zone, for steady-state, also (Figure 5). The molten pool is painted dark gray color. The heat affected zone width corresponds to maximum size of light gray color area along y axis defined by temperature $T = 900\,K$.

Figure 5. Influence of source power density on shape and size of molten pool and heat affected zone at the moment of steady state quasi-stationary regime. $q_{m0,B} = 0.7\ 1/s; q_{m0,Ti} = 2.1\ 1/s$.

Heterogeneous character the molten pool is connected with chemical reactions and mixing. The concentration distributions $C_i(x,0)$ at different times t and for different values q_0 of power density source are shown in Figure 6. Obviously, the higher source power density, the larger number of particles dissolving in the molten pool. Figure 6b allows estimate composition of formed surface layer.

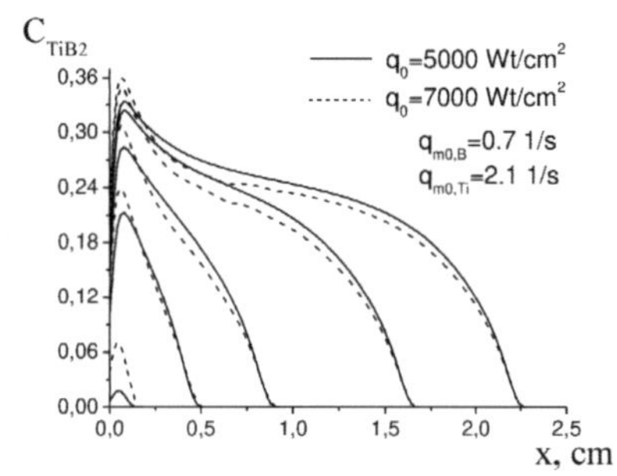

a)

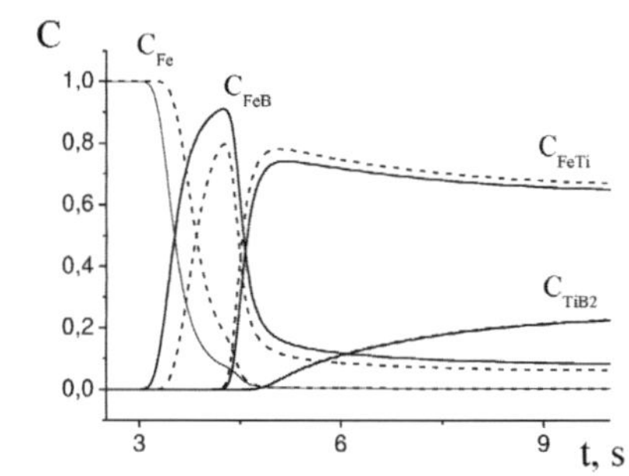

b)

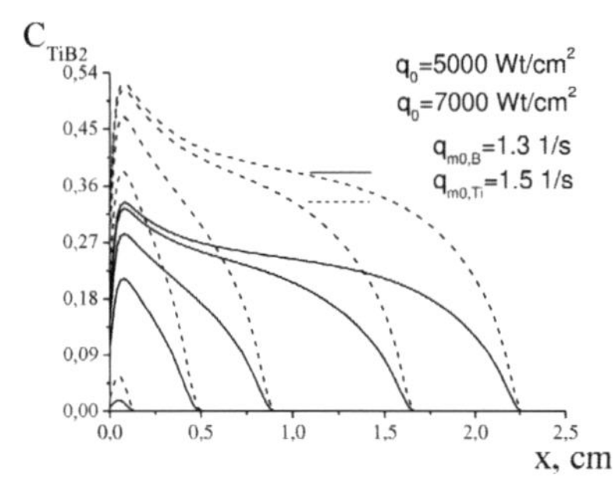

c)

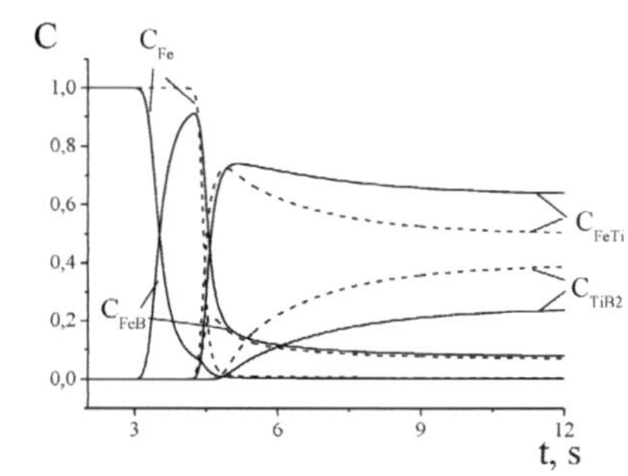

d)

Figure 6. a), c) The distribution of the relative TiB_2 mass concentration along Ox axis at various times t; b), d) dependence of the relative mass concentrations of iron and forming compounds over time for different density of the particle flux. $V = 0.2$ cm/s.

6. Conclusion

1. The electron beam surfacing of thermoreactive Fe-B-Ti powders allows producing composite coatings on the basis of refractory titanium borides synthesized under the electron beam in vacuum. Such coatings exhibit high abrasive wear resistance and low wear rate in dry friction.
2. The granulometric composition of the fused powder mixtures strongly influences the physico-chemical processes occurring in the molten bath under the electron beam. It defines the coating properties that depend on the structure, proportion of boride phases and the metal binder, and the degree of solid solution hardening.
3. The model of the technological process of electron beam treatment of the material surface with modifying particles titanium and boron with accounting of formation of chemical compounds is supposed. All formal kinetic parameters included in the model are estimated. It is shown that the phase

and chemical composition of formed coating strongly depends on both the relations of components (fraction of particles) and the parameters characterizing the rate of heating (rate of movement of the source, flow density of energy and particles). The dependence of the coating composition on the beam power density and the density of the particle flux is illustrated. Similar dependence was observed in the experimental investigations performed on real systems [3, 10].

Acknowledgments

The work was carried out in the framework of the Fundamental Research Program of the State Academies of Sciences for 2013–2020. Project 23.2.1.

References

[1] Paton B E 1973 *Electron beam coating and refining of metals and* alloys (Kiev: Naukova Dumka)
[2] Galchenko N K, Belyuk S I, Panin V E *et al* 2002 *Fiz. Khim. Obrab. Materialov* **4** 68
[3] Kolesnikova K A, Galchenko N K 2006 *Physical Mesomechanics* 9 165–168
[4] Kryukova O N, Knyazeva A G 2006 *Physical Mesomechanics* 9 65–68
[5] Knyazeva A G, Pobol I L, Gordienko A I, Demidov V N, Kryukova O N, Oleschuk I G 2007 *Physical Mesomechanics* 10 **3-4** 207–220
[6] Kryukova O N 2006 *Bulletin TPU* **6** 120 (in Russian)
[7] Frolov V F 1998 *Theoretical Foundations of Chemical Engineering* 32 **4** 398
[8] Kryukova O N, Knyazeva A G 2014 *Russian Physics Journal* 57 **9/3** 113 (in Russian).
[9] Kryukova O N, Knyazeva A G 2007 *Journal of Applied Mechanics and Technical Physics* 48 **1** 109–118
[10] Galchenko N K, Belyuk S I, Kolesnikova K A, Galchenko V G 2010 *Deformation and fracture of materials* **4** 25

IOP Conf. Series: Materials Science and Engineering **140** (2016) 012012 doi:10.1088/1757-899X/140/1/012012

Structures Formation on the Y–TZP–Al_2O_3 Ceramic Composites Surface

Sergei Kulkov[1,2,3,a], Irina Sevostyanova[1], Tatiana Sablina[1], Svetlana Buyakova[1,2,3], Artem Pshenichnyy[3] and Nickolai Savchenko[1,2,3]

[1]Institute of Strength Physics and Materials Science SB RAS, 2/4, pr. Akademicheskii, Tomsk, 634055, Russia

[2] Tomsk Polytechnic University, 30, Lenin Avenue, Tomsk, 634050, Russia

[3] Tomsk State University, Tomsk, 36, Lenin Avenue, Tomsk, 634050, Russia

[a] Corresponding author: *kulkov@ms.tsc.ru*

Abstract. The paper discusses the structure of Y–TZP–Al_2O_3 ceramics produced from nanopowders and friction surface, wear resistance, friction coefficient of Y–TZP–Al_2O_3 composites rubbed against a steel disk counterface at a pressure of 5 MPa in a range of sliding speeds from 0.2 to 47 m/s. Analysis by X-ray diffraction, scanning electron microscopy showed that the high wear resistance of Y–TZP–Al_2O_3 composites at high sliding speeds is due to high-temperature phase transitions and protective film formation on the friction surface.

1. Introduction

Additive technologies, or technologies of layer-by-layer synthesis is one of the most dynamically developing directions of "digital" production. Zirconia-based ceramics produced from nanopowders is considered one of the most promising materials for additive technologies thanks to its ability to widely change the mechanical properties as a result of structural changes [1]. Particularly promising is the production by means additive technologies of zirconia high speed bearings.

The ZrO_2 tetragonal phase transformation into the monocline phase under the effect of applied stresses is the main toughening mechanism of the Y-TZP-based ceramics [1]. One specific feature of the above materials is that the temperature rise reduces the contribution of the transformation toughening hardening mechanism to their mechanical properties. Especially strongly reduced strength properties in Y-TZP ceramics without additives of other oxides.

The Y-TZP–Al_2O_3 composites are free of this drawback because the Al_2O_3 particles act as a hardening additive. The Y-TZP–Al_2O_3 composites are widely popular because of their relatively high thermal stability [1]. As follows from the data presented in the literature, usually by friction tests in a pair of Y-TZP/ceramic end even at speeds above 1 m/s due to the catastrophic destruction of the friction surface [2,3].

The upper limit of the sliding speeds can be raised to 10–40 m/s if the steel serves as the counterbody [4, 5]. This is caused by the transfer processes of the metal and its oxides to the ceramic surface, producing a so-called transfer layer fulfilling the protective function and promoting the wear resistance of the ceramics at fast sliding speeds [5]. However, the published data on the friction of zirconium-dioxide ceramics are not sufficient to draw conclusions on the processes evolving on the friction surfaces or the wear mechanisms within a broad range of speeds and loads in order to choose the optimum structure of the material and expand the potential of its practical application.

The aim of the current work is to study the structure of friction surface and tribological characteristics of the Y-TZP–Al_2O_3 ceramics in dry friction steel interfaces within a broad range of sliding speeds (between 0.1 and 47 m/s).

2. Materials and experimental procedure

The Zr, Y, and Al aqueous salt solutions served as the original materials for obtaining transformation hardened ceramic composites; the salts were turned into powders in the high-frequency plasma discharge. They had the following composition: 80 wt % ZrO_2 (3 mol % Y_2O_3) +20 wt % Al_2O_3. X-ray phase analysis revealed that powder contained a mixture of three phases: the ZrO_2 tetragonal and Al_2O_3 γ- and ε- modifications. The Y-TZP–Al_2O_3 powders consisted of hollow spheres and their fragments (Figure 1a). The walls of the hollow spheres were polycrystals with a grain size no greater than 20 nm.

The ceramic specimens were compacted in a metallic press mould and then sintered in vacuum at 1700°C for 3 hours. The microstructure of the sintered ceramic shown in Figure 1b.

X-ray phase analysis revealed that after sintering produced tetragonal (T) and cubic (K) zirconium-dioxide phases and α-phases Al_2O_3 in the ceramics. Mechanical properties of sintered composites are shown in Table 1. The sintered ceramics had high hardness, high bending strength and fracture toughness. After sintering, metallographic analysis revealed the residual porosity of the ceramics equal to 5 vol %.

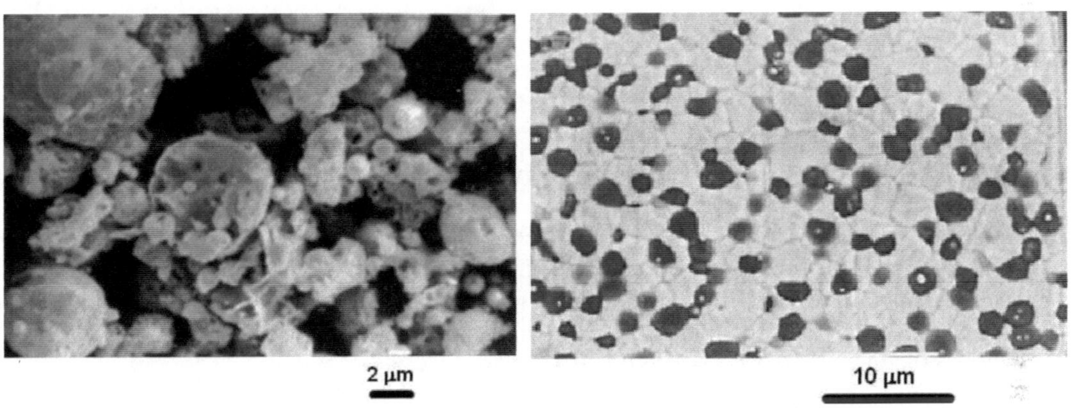

a) b)

Figure 1. Particle morphology of Y-TZP–Al_2O_3 powders obtained by the decomposition of salts in the plasma of a high frequency discharge (a). Microstructure of Y-TZP–Al_2O_3 specimens after sintering. The surface of the sintered Y-TZP–Al_2O_3 specimen was thermally etched at 1500°C for 10 min (b).

Table 1. - Mechanical properties of the composites

Relative density	Hv, GPa	Bending Strength, MPa	K_{Ic}, MPa*m$^{1/2}$
0.95	11.4	900	9

The pin-on-disk pairs were tested in dry friction on a UMT-1 unit with the sliding speed increased in steps. The counterbody was a disk from cast steel with the martensite structure (HRC 63–65), $M_{12}C$ carbides, and minimal residual austenite. The specimens of the ceramics in question had rectangular sectional area 60 mm^2. The test pressure was 5 MPa and the sliding speeds ranged between 0.1 to 47 m/s; the test duration was selected so that the friction path would be 2 km at any speed.

3. Results and discussion

The tribotests carried out in the range of sliding speeds from 0.2 to 47 m/s at a pressure of 5 MPa showed that with increasing the sliding speed wear rate of Y-TZP-Al_2O_3 ceramics first increases and then decreases (Figure 2a). The friction coefficient of the ceramics decreases from 0.5–0.8 at low speeds to 0.15–0.20 at speeds of 25–47 m/s (Figure 2b).

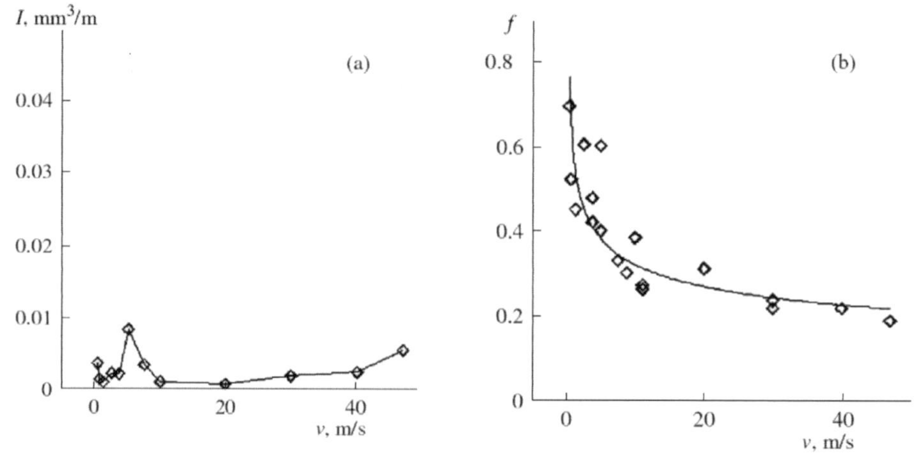

Figure 2. Wear rate (a) and friction coefficient (b) of Y-TZP–Al_2O_3 ceramics as a function of sliding speed on steel.

Scanning-electron and optical microscopy demonstrated that the ceramic surfaces acquired a grooved relief typical of the abrasive wear mechanism after friction within the speed range up to 1 m/s (Figure 3a). The worn surfaces manifested extensive areas with traces of delamination and spalling of the material at speeds >3 m/s (Figure 3b) , while the ceramic surfaces had a rather smooth relief at speeds >6 m/s (Figure 3c). It is particularly noteworthy that the Y-TZP–Al_2O_3 friction surfaces contain a cracking mesh that divides them into individual fragments (Figure 3b,c). Measurement of the spacing between the cracks in the sliding direction has revealed that their size distribution has the pattern close to a normal distribution with a distinct maximum.

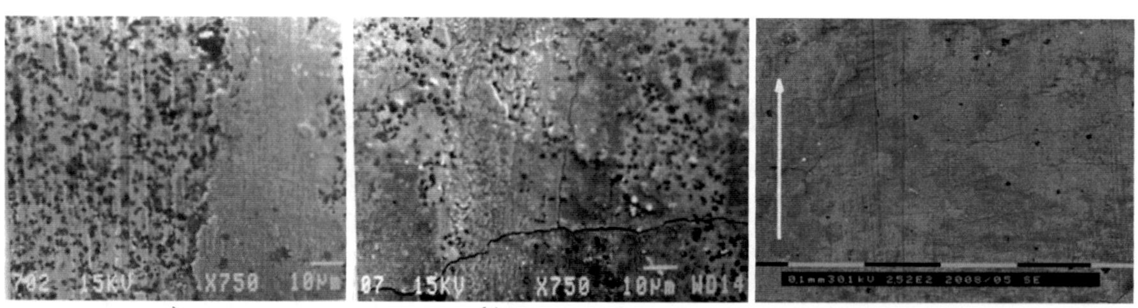

a) b) c)

Figure 3. SEM images of friction surface of Y-TZP–Al_2O_3 ceramics after tests at: v= 0.9 m/s (a); v=4.7 m/s (b); v= 47 m/s (c). Arrow indicate sliding direction.

According to the X-ray phase analysis of the friction surfaces of the Y-TZP–Al_2O_3 ceramics, there are three (monocline, cubic, and tetragonal) crystalline modifications, depending on the sliding speed. It should be noted that the monocline phase does not exist in the original ceramics, while it exists only in the amount of 7–15% by volume after tests at slow sliding speeds, for example up to 11.1 m/s of the Y-TZP–Al_2O_3 ceramics.

The appearance of a monoclinic phase on the friction surface of Y-TZP-Al_2O_3 evidence of martensitic transformation of the metastable tetragonal phase to the stable monoclinic phase. Reduction of the monocline zirconium-dioxide phase as the sliding speed accelerates can be attributed to growth in the contact-zone temperature and reduction of the thermodynamic stimulus for the tetragonal monocline transformation [1]. The concentration of the K-phase typical of the original condition of the ceramics (10 vol. %) remains unchanged in the two materials up to the speed 20 m/s. If the sliding continues to accelerate, its concentration in the specimens grows after the tests to ≈15 vol %. The emergence of the cubic phase on the surfaces of the specimens at the maximum sliding speeds is likely due to the fact that a part of the tetragonal phase undergoes diffusion and transforms into the cubic phase. This phase is favored by the elevated temperatures in the contact zone, assessed according to [6] to be on the order of 2000°C at the speed of 10 m/s. Optical and scanning-electron microscopy reveals that the Y-TZP–Al_2O_3 friction surfaces at sliding speed above 6 m/s become relatively evenly covered with a transfer layer. The transfer layer at slow and moderate speeds is distributed extremely irregularly over the surface, leaving large areas completely uncovered.

It has previously been shown [5] that of the subsurface regions of the Y-TZP ceramics reveals a region of material after the tests within the speed range from 0.1 to 4 m/s that underlies the fine layer (1–2 μm) and has a subsurface layer with the submicrocrystalline structure in which the shapes of the grains are considerably changed in the sliding direction. This zone is thickest (about 10 μm) at moderate sliding speeds (≈4 m/s), i.e., when the wear rate of the ceramics matters most. The differently shaped grains disappear in the Y-TZP ceramic specimens as soon as the speed exceeds 6 m/s; the grains remain the uniaxial, as they originally were. The differently shaped grains are not observed in the Y-TZP–Al_2O_3 ceramics throughout the whole range of speeds (Figure 4), including at the moderate sliding speeds when the material undergoes intensive wear (Figure 4a).

a) b)

Figure 4. SEM images of structure of subsurface regions of Y-TZP–Al_2O_3 ceramics after friction at v = 4.3 m/s (a) and v = 11.1 m/s (b). Arrows indicate sliding direction.

The wear processes evolving at speeds 2–4 m/s correspond to high-temperature adhesive - notwithstanding the common tribological behavior of the Y-TZP [5] and Y-TZP–Al_2O_3 ceramics, the latter material containing the Al_2O_3 particles manifests stronger wear resistance at moderate (3–4 m/s) sliding speeds. In all probability, the structure of this ceramic composite with hardening aluminum-oxide particles has a stronger surface load-bearing capacity during friction when the continuous transfer layer has not yet formed on the ceramic surface.

4. Summary

The accomplished experiments have demonstrated that the Y-TZP–Al_2O_3 ceramic specimens sintered from nanopowders possess high wear resistance at high sliding speeds close to 47 m/s, irrespective of the elevated temperature in the contact zone resulting in high-temperature phase transitions. A broad range of sliding speeds has been identified (6–47 m/s) within which the ceramics undergo virtually no wear because a quasiliquid film appears and it coats regularly their surface.

Acknowledgement
The work was financed in part by the Ministry of Education and Science of the Russian Federation in the framework of Agreement No.14.575.21.0040 (RFMEFI57514X0040).

References

[1] Nettleship L and Stevens R 1987 *Int. J. High Technology Ceramics* **3** 1–32
[2] Lee S W, Hsu S H and Shen M C 1993 *J. Amer. Ceram. Soc.* **76/8** 1937–47
[3] He Y J, Winnubst A J A, Burggraaf A J, et al., 1995 *J. Amer. Cream. Soc.***78/2** 3090–96.
[4] Kulkov S and Savchenko N 2008 *Epitoanyag* **3** 62 - 64
[5] Savchenko N L, Pyatova K M and Kul'kov S N 2008 *Vestn. Tomsk. gos. un–ta. Mat. Mekh.* **1** 84–89
[6] Lim S C and Ashby M F 1987 *Acta Metallurgica* **35** 1–24

Use of additive technologies for practical working with complex models for foundry technologies

E Olkhovik, A A Butsanets and A A Ageeva
Research and Innovation Department, Admiral Makarov State University of Maritime and Inland Shipping, Saint-Petersburg, 198035, Russia

E-mail: olhovick@gmail.com

Abstract. The article presents the results of research of additive technology (3D printing) application for developing a geometrically complex model of castings parts. Investment casting is well known and widely used technology for the production of complex parts. The work proposes the use of a 3D printing technology for manufacturing models parts, which are removed by thermal destruction. Traditional methods of equipment production for investment casting involve the use of manual labor which has problems with dimensional accuracy, and CNC technology which is less used. Such scheme is low productive and demands considerable time. We have offered an alternative method which consists in printing the main knots using a 3D printer (PLA and ABS) with a subsequent production of castings models from them. In this article, the main technological methods are considered and their problems are discussed. The dimensional accuracy of models in comparison with investment casting technology is considered as the main aspect.

1. Introduction

Additive technology (AT) has been used in tasks aimed at creating material models in no time. At the same time, a distinctive feature of AT is that there is no need in using additional or special tools (equipment). Metal including steel details with complex geometry and surface shape are produced mainly by casting methods using investment casting or lost foam casting for burned-out models because traditional die casting with the use of sand mold has limited capabilities [1-2]. The major drawback of investment casting is high labor consumption due to use of manual work and low dimensional accuracy of castings, since the material of the model inherently has low durability and rigidity, especially taking into account the large number of technological operations it is subjected to [3].

The main objective of the work was to study the possibility of AT application for fast manufacture of high-precision models with the subsequent manufacture of metal castings. Development of die casting is complicated and cyclic process, where it is necessary to develop a large number of technological solutions, both for the whole casting and its separate parts. In practice this means the production of a large number of variants of the model and auxiliaries, which are die sets, cores, gating systems and risers. Obviously, special equipment, highly skilled personnel and considerable time expenditures are required for this reason. In this work, we have suggested to put aside intermediate industrial equipment at the stage of die casting development due to AT application, namely manufacture of models using a 3D printer.

2. Materials and Methods

The main area of work was studying dimensional tolerance, quality of experimental castings surface and thermodestruction of modeling material, because the distinctive feature of the developed method is that the casting model is not preliminarily removed previously (unlike the Investment Casting method), but it is burned out by the melted metal. The defining feature of this research is the use of PLA and ABS material as the modeling material. The studies of properties of other materials for model production are presented in [2, 4-6]. Low melting temperature, biodegradability, stability during production of prepared models and easy machining completely meets the requirement to the produced models. PLA and ABS plastic has amorphous structure, which is an advantage for receiving smooth surface or surface of complex relief.

The experiments employed a typical example of 3 or 4 walls with different thickness combined into a single unit, model of a cast body of a stop valve (Figure 1a). Such object is the most difficult and widely spread at the stage of formulation of requirements for determining the thickness of walls depending on the feeding system and defining the design of radial blends between the walls. The design of such difficult unit, as a rule, is defined only by direct experiments, since the thickness of walls and radii of blend between them cannot be calculated both by analytical methods and by computer modeling methods.

The tasks of research included the development of computer parametrical model of casting (Figures 1a and 1b) with a capability of fast change of geometrical dimensions [7], printing of the model using a 3D printer, coating it by a refractory layer, study of burning properties of ABS and PLA for forecasting casting by molten metal. The models were printed using Makerbot Replicator 2 printer by Fused Deposition Modeling (FDM), i.e. layer-by-layer covering of the melted construction material by an extruder. According to ASTM International classification, this method belongs to Material Extrusion. Today it is the most affordable consumer technology, which is the basis for currently produced industrial and professional machines. ReplicatorG (open source 3D printing program) was used as the slicer (the software for converting of three-dimensional model into a set of movement commands for the printing head).

The desired result of printing was received with the following settings: layer height is 0.15 mm, object infill is 10%, printing temperature for PLA is 230 °C and printing temperature for ABS is 250 °C. Heating of table is 100 °C. Printing of a half of the model for both types of plastic took about 180 minutes.

Figure 1. Initial design of cast valve body

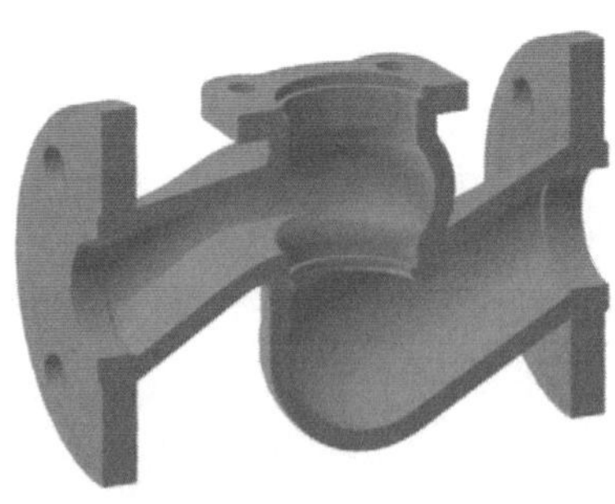

Figure 2. Design of cast valve body for 3D printing

In order to form an internal cavity, the body was divided into two parts (Figures. 2, 3 and 4) diametrically to the central plane of symmetry. Parting planes of half-models were subjected to machining by an emery paper P160 using the grinder.

The printed 3D model from PLA plastic was pre-treated by chloride methylene (dichloromethane) for the sake of high-quality surface; the model from ABS plastic was not chemically treated.

Figure 3. Inner surface of ABS-model after removing the pattern-supporting layer

Figure 4. Model after machining the parting plane

3. Application of the refractory layer

The refractory layer was applied in two stages. The first stage was the filling of the internal space; the second stage was the creation of the shell. The internal cavity was filled with mix on the basis of fine quartz sand and plaster binding in equal proportions, water was added for ensuring plasticity of the mix. The resulting solution was pressed through one hole before complete filling, which allowed reducing the probability of emergence of air cavities to minimum (Figure 5). In 24 hours the model was disassembled, the core was completely hardened. Then, the two halves of the model were put together and bound with a special thread.

The external refractory layer was deposited by method of dipping into the solution on the basis of quartz sand and organic binding in two layers with intermediate drying of 12 hours. Drying of the refractory layer was carried out in a vacuum oven with blowing by warm air (35-40°C). PLA plastic has good adhesion with the refractory layer; there was no peeling or destruction of the refractory layer (Figure 6).

Figure 5. Rammed core in the model

Figure 6. Model with the deposited refractory coating after 12 hours

4. Studying thermodestruction of model material (ABS and PLA) and its properties

Important feature of the developed technology is the possibility of direct removal of ABS and PLA model due to impact of high temperature of the melted metal. The initial burning temperature of both plastics is ~395 °C; at the same time, the process has the maximum intensity at the temperature of more than 425 °C, and it is accompanied by release of gases. Products of thermodestruction of PLA plastic are considered harmless; ABS plastic partially emits toxic agents. In this work, the rates of model material burning were investigated, because slow gasification can result in unsatisfactory filling of a form by liquid metal and, as a result, to the formation of gas sinks or gas porosity in future casting due to dissolution of gas in metal. The technique of experiment consisted in intense temperature impact on ABS and PLA plastic with the measurement of burning rate and identification of residual products. The standard wire for printing with the diameter of 1.75 mm was used. Thermodestruction rate was investigated by kinetic movement of the burning edge under conditions of steady burning of the material. Speed of burning of the PLA thread was 2.25 mm/sec, and for the ABS thread it was 3.33 mm/sec. With the increase in temperature, the speed of burning will proportionally increase, but from the received measurements, it is possible to make a conclusion about low speed of thermodestruction of plastics.

5. Visual and instrumental assessment of chosen parameters

Layer-by-layer material deposition causes uneven cooling of layers, which leads to distortion of parts. When the printing of models ended, the edges of PLA model were detached from the table (the shape distortion is obvious), and the ABS model completely adhered to the table. ABS plastic turned out to have better adhesion. The deformation of substrate layers was noticed, which indicated the distortion. This was confirmed at the interface of model parts. Both plastics required machining of the parting plane.

The linear dimensions of parts increased without obvious regularity, probably, according to the settings of the software (slicer) when converting 3D-drawing into the commands for the printer (Table 1). The holes are not straight and are out of plane. The printed holes are often oval. Visual control showed that the holes printed on the axis of a circle mainly deviate from planeness, and the holes printed perpendicularly to the axis deviate from straightness.

Table 1. Deviations of the actual sizes of model from sizes given in the drawing in percent

	drawing	model (PLA)	model (ABS)
Length	100	+1.010	+1.016
Flange thickness	100	+1.048	+1.086
Flange diameter	100	+1,004	+1.013
Thickness of the cover flange	100	+1.119	+1.048

According to the standard "Geometrical Product Specifications (GPS) — Dimensional and geometrical tolerances for moulded parts — Part 3: General dimensional and geometrical tolerances and machining allowances for castings (ISO 8062-3:2007)", the parts from PLA and ABS had dimensional casting tolerance grades DCTG7. According to the INTERNATIONAL STANDARD ISO 8062 Second edition 1994-04-01 Castings — System of dimensional tolerances and machining allowances] casting tolerances — the grade was CT7, and CT8 for the ABS detail.

6. Conclusion

In this research, it was established that application of additive technologies can be used for fast production of castings models, including those with irregular shape with a sufficient accuracy.

When using PLA plastic, the model has bigger tendency to distortion of initial layers, as compared with ABS. It is possible to prevent such effect by choosing optimal parameters of table heating, to keep the printer working chamber warm, to improve adhesion by applying special structures to the table before the printing.

When placing the part on the table, it is important to consider that support will be necessary for all protruding elements. Therefore, it is recommended to place the part on the table with the wide element facing down. The larger the interface surface between the part and the table, the lesser the probability of separation.

If there are requirements to dimensional accuracy, the manufactured models should be checked for deviations of the size and shape from set values. Special attention should be paid to the geometrical sizes of holes and their relative arrangement.

The use of different software (slicers) can lead to the different sizes of the model.

Too dense refractory solution can crack when drying. Obviously, it is connected with adhesive properties of PLA plastic.

The problem of low thermodestruction speed of plastics can be solved by increase in temperature of filling by 10-15 °C, use of an open sprue channel of increased diameter (up to 20%) in cross-sectional area, increase of form filling time, application of open risers for removal of burn products.

It is necessary to perform works in well aired rooms. PLA combustion occurs rather slowly, plastic transforms into liquid, it is not fully burned out without release of acrid substances. ABS burns rather actively, transits into liquid phase and burns, emitting acrid substances and particles of soot in air.

References

[1] Chhabra M and Singh R 2011 Rapid casting solutions: a review *Rapid Prototyping Journal.* **17.5** 328–50

[2] Mendonsa C and Shenoy V D 2014 Additive Manufacturing Technique in Pattern making for Metal Casting using Fused Filament Fabrication Printer *Journal of Basic and Applied Engineering Research* **1.1** 10–13

[3] Cheah C M, Chua C K, Lee C W and Totong K 2005 Rapid prototyping and tooling techniques: a review of applications for rapid investment casting *The International Journal of Advanced Manufacturing Technology* **25.3-4** 308–20

[4] Marwah O M F, Sharif S, Zainol M A, Ibrahim M and Mohamad E J 2014 3D Printer Patterns Evaluation for Direct Investment Casting *Applied Mechanics and Materials* **465–466** 1400–1403 (DOI: 10.4028/www.scientific.net/AMM.465-466.1400)

[5] Gill S S and Kaplas M 2011 Efficacy of powder-based three-dimensional printing (3DP) technologies for rapid casting of light alloys *The International Journal of Advanced Manufacturing Technology* **52.1–4** 53–64

[6] Meisel N A, Williams C B and Druschitz A 2012 Lightweight metal cellular structures via indirect 3D printing and casting *Proceedings of the International Solid Freeform Fabrication Symposium* 162–76.

[7] Ol'khovik E O, Butsanets A A and Ageeva A A 2015 Use of the distributed computing at the castings solidification simulation *International Conference on Mechanical Engineering, Automation and Control Systems (MEACS)* pp. 1–4 (DOI: 10.1109/MEACS.2015.7414905)

[8] Ol'khovik E O, Reznik U A 2014 Development of laser welding technology of bellows assemblies for ship pipeline valves *Vestnik Gosudarstvennogo Universiteta Morskogo i Rechnogo Flota imeni admirala S O Makarova* **25** 119–122

Process Conditions of Forming the Surface Layer of Aluminum Powder Product by Layer-by-layer Laser Sintering

N A Saprykina[a], A A Saprykin, E A Ibragimov, D A Arkhipova
Yurga Institute of Technology, Tomsk Polytechnic University, Tomsk, Russia

E-mail: [a] saprikina@tpu.ru

Abstract. The paper presents data on state of the art in selective laser sintering of products. Layer-by-layer sintering is shown to be a future-oriented technology, making it possible to synthesize products of metal powder materials. Factors, influencing the quality of a sintered product, are revealed in the paper. It presents outcomes of experiments, focused on the dependence of surface layer thickness of sintered aluminum powder PA-4 on laser processing conditions. Basic factors, influencing the quality of a sintered surface layer include laser power, speeds of scanning and moving the laser beam on the layer of powder. Thickness of the sintered layer varies from 0.74 to 1.55 mm, as the result of changing the laser processing conditions.

1. Introduction

The present day economy makes use of layer-by-layer laser sintering technologies in order to reduce production time and manufacture competitive products. These technologies enable synthesizing products of any configuration from metal powder materials with a preset accuracy on the base of a 3D CAD-model, requiring no preliminary costs for tooling preparation. These technologies have an extra advantage; it is a considerable material economy when manufacturing products of complex configuration [1-4].

A 3D CAD-model is separated in layers when manufacturing a product by layer-by-layer laser sintering. Some powder is fed into the area of sintering; a layer coinciding with the section of a product is formed via moving a laser beam on the section area in preset conditions. This process is completed as soon as a product is made. Layer-by-layer laser sintering has one more advantage: a powder material is used efficiently, since it is possible to reuse non-sintered powder.

Alongside with advantages additive technologies face some restrictions regarding high cost of the equipment and powder materials, as well as their assortment. Poor quality of formed surface layer of products, insufficient mechanical strength and probable delaminating are burning issues highlighted by the authors [7, 8, 9].

Further development of layer-by-layer laser sintering is possible provided that new powder materials are applied, functionally graded structures are made [10]. Synthesizing a product of special quality, with certain mechanic strength from metal powder materials requires taking into consideration factors, influencing the quality of sintered surface and its inner structure:

- thermal impact (surface bulging, cracking, delaminating, and melting) [11];

- process conditions of layer deposition (compaction, speed of deposition, layer displacement when depositing a powder) [12];
- process conditions of sintering (density of irradiation power, impulse energy, speed of laser beam moving, diameter of a laser spot) [13];
- material of powder (heat-transfer properties, physical and chemical characteristics, dimensions and configuration of powder) [14].

Factors mentioned above emphasize that a functional product can be manufactured in a certain range of laser impact only, which is specified experimentally for each powder material, and accuracy of a prototype depends mainly on the sintered layer thickness.

This paper presents outcomes of experiments focused on the influence of laser processing conditions on the surface layer thickness of aluminum powder PA-4 produced by layer-by-layer laser sintering.

Aluminum powder is widely used in engineering due to its good physical properties, but industry making the most use of it is aircraft construction. Powder PA-4 is produced from primary aluminum at melting temperature 660 °C. It consists of active aluminum (not less than 98%), ferrum (0.35%), silicon (0.4%), and copper (0.02%).

2. Results and Discussion

The effect of sintering process conditions (laser beam speed V, laser power P, and scanning step S) on the surface layer thickness of the sintered aluminum powder PA-4 is studied in this paper. Experiments were carried out using a specially developed machine of layer-by-layer laser sintering, which supports adjustability of all process conditions of sintering. This is a laser beam machine to form the surface of products with a complex physical configuration, which consists of ytterbium fiber laser LK – 100 – V (wavelength 1.07 μm), a three-dimensional table, a PC, a CNC system, and specially developed software. A stepping motor is used for moving in coordinates X, Y, Z [15].

Since layer-by-layer laser sintering is used for product manufacturing, we should know the thickness of a sintered layer when separating a CAD-model in layers. The same value is also needed when manufacturing a product. Unless a sintered layer is thick enough, it is difficult to deposit the top powder layer as the bottom one gets destructed. Increasing the thickness of powder to be deposited leads to delaminating of a sintered product. Layer thickness is important for the quality of a steplike side face. The height of each step approximates to the thickness of a sintered layer. Process conditions of sintering are to be assigned properly, so the product meets standard dimensions within necessary tolerance of quality. A method providing high accuracy of separating in layers a virtual 3D model of a product is proposed in patent of invention 2262741 RF [16]. Thickness of layers can't exceed a nominal profile tolerance range of a model surface; a nominal profile of the model surface coincides with the midline of cross layers, so that is an actual profile. Thickness of the layer is assigned in the range:

$$Zmin \leq Z \leq Zmax, \qquad (1)$$

where Zmin and Zmax – minimal and maximal thickness of a layer, respectively, provided by technological equipment.

Assignment of appropriate process conditions of sintering makes it possible to produce a three-dimensional product with a complex physical configuration [17].

Samples of a sintered 20 mm long and 10 mm wide layer were produced for the purposes of experiments. Laser beam movement is shown in Figure 1.

Process conditions of laser sintering tested in experiments were pre-determined in exploratory experiments. Appropriate conditions for manufacturing a sample of powder PA-4 with standard mechanical strength are as follows: P = (10-20) W, V = (1000-3000) mm / min, S = (0.1-0.2) mm. The geometry of the sintered surface was analyzed by means of a specially developed technique based on Digital Microscope toolkit.

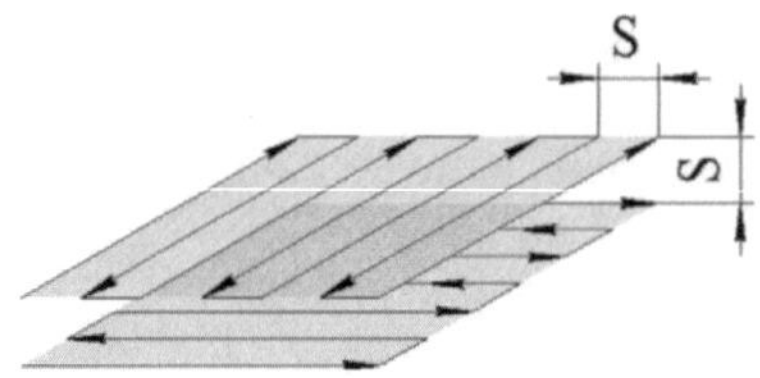

Figure 1. Scanning of layers

Figures 2-3 show, how the laser power influences on the quality of a sintered surface layer. Power varying from 10 to 20 W at constant speed V = 1000 mm / min, sintering temperature t = 26 °C and scanning step S = 0.2 mm cause the increase in thickness of a sintered layer from 0.765 to 1.55 mm. In Figure 3 one can see more coagulated particles because of the increased density of energy.

Figure 2. Physical configuration of the sintered PA-4 surface (x2), process conditions of sintering: V=1000 mm/min, t=26 °C, S=0.2 mm, P=10 W.

Figure 3. Physical configuration of the sintered PA-4 surface (x2), process conditions of sintering: V=1000 mm/min, t=26 °C, S=0.2 mm, P=20 W.

Mechanical strength of samples gets decreased as the result of reducing the power to 10 W. Assigning the following process conditions: power 20 W, V = 3000 mm / min, t = 26 °C, S = 0.2 mm is the reason that the sintered sample has thickness of 1.33 mm (Figure 5).

Figure 4. Physical configuration of the sintered PA-4 surface (x2), process conditions of sintering: V=3000 mm/min, t=26 °C, S=0.2 mm, P=10 W.

Figure 5. Physical configuration of the sintered PA-4 surface (x2), process conditions of sintering: V=3000 mm/min, t=26 °C, S=0.2 mm, P=20 W.

Figures 6-7 show, how the surface of sintered powder PA-4 changes in dependence on the speed of laser beam movement. The increase in the speed of laser beam movement from 1000 to 3000 mm / min, at P = 20 W, t = 26 °C, S = 0.2 mm leads to the decrease of thickness in the sintered layer from 1.55 to 1.33 mm, (Figures 6-7).

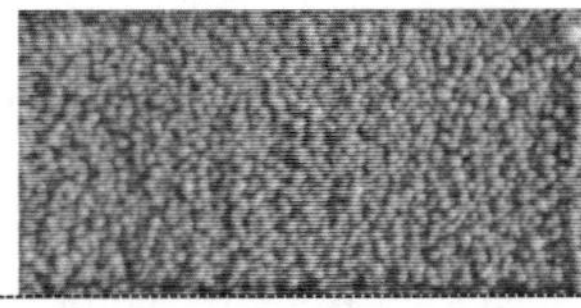

Figure 6. Physical configuration of the sintered PA-4 surface (x2), process conditions of sintering: V=1000 mm/min, t=26 °C, S=0.2 mm, P=20 W.

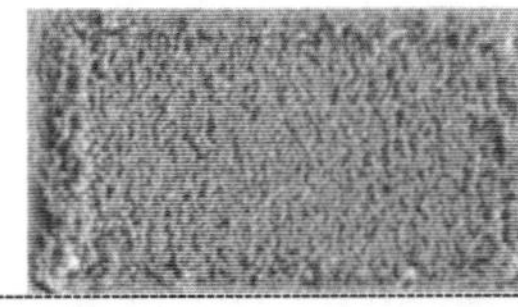

Figure 7. Physical configuration of the sintered PA-4 surface (x2), process conditions of sintering: V=3000 mm/min, t=26 °C, S=0.2 mm, P=20 W.

Comparing the samples one can see that increasing the speed of laser beam movement can cause the decrease in the thickness of sintered layer.

In Figures 8-11 one can see physical configuration of sintered PA-4 surface regarding the scanning step.

Increasing the scanning step from 0.1 to 0.2 mm (process conditions of sintering: P = 10 W, t = 26 °C, V = 1000 mm / min (Figures 8-9) results in the decrease of sintered layer thickness from 0.74 to 0.765 mm.

Figure 8. Physical configuration of sintered PA-4 surface (x2), process conditions of sintering: V=1000 mm/min, t=26 °C, S=0.1 mm, P=10 W.

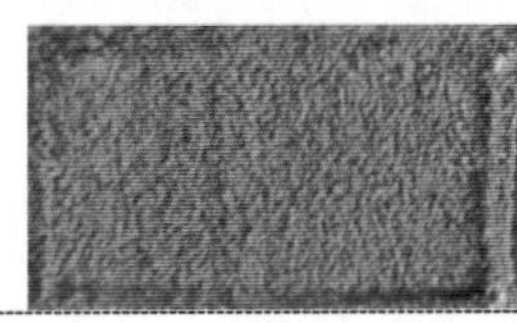

Figure 9. Physical configuration of sintered PA-4 surface (x2), process conditions of sintering: V=1000 mm/min, t=26 °C, S=0.2 mm, P=10 W.

In samples obtained in process conditions of sintering: P = 15 W, t = 26 °C, V = 3000 mm / min, scanning step varies from 0.1 to 0.2 mm (Figures 9-10), as the results, thickness of the sintered layer increases from 1.0 to 1.2 mm.

Figure 10. Physical configuration of the sintered PA-4 surface (x2), process conditions of sintering: V=3000 mm/min, t=26 °C, S=0.1 mm, P=15 W.

Figure 11. Physical configuration of the sintered PA-4 surface (x2), process conditions of sintering: V=1000 mm/min, t=26 °C, S=0.2 mm, P=15 W.

Increasing the scanning step causes the growth of the thickness of sintered PA-4 layer. No visible defects are detected on the sintered samples.

Our studies suggest a significant influence of power on quality of the sintered surface layer. Unless power is assigned correctly samples are deformed. The speed of a laser beam movement is also important for the quality of the surface layer. In some cases, to form a surface one needs to increase the speed for reducing the thickness of the sintered layer. Scanning step is not so relevant for the quality of the sintered surface layer. When sintering aluminum powder increased scanning step causes the growth of the sintered layer thickness. This allows to determine the number of layers in the finished product and to produce a three-dimensional product of any geometrical shape. In Figure 12 a three-dimensional sample, consisting of 10 layers is shown. The increment thickness of layers is 200 μm, sintering was carried out in the shielding atmosphere of argon.

Figure 12. Physical configuration of a three-dimensional sample (x2). Process conditions of sintering: V=1000 mm/min, t=26 °C, S=0.1 mm, P =60 W, increment thickness of layers 200 μm, sintering in argon

The surface of the sample is rough, gapless grooves are seen, in particular along the path of laser beam movement. Typical beads, coagulated from the melt, are 1-1.5 mm; that differs by a factor of ten from the initial size of mixed powder composites. This fact makes it obvious laser power is sufficient for coagulation of particles of the melt into bigger formations. A developed inner porosity has a positive effect on adhesion of layers.

3. Conclusion

The authors have pointed at the possibility of wide range changing the sintered surface layer due to variation of process conditions of laser sintering. Increasing the speed from 1000 to 3000 mm/min results in augmentation of layer thickness by 16 % in conditions: P=20 V, S=0.2 mm, t=26 °C. Increasing the power from 10 to 20 V results in augmentation of sintered layer thickness by 200 %, in conditions: V=1000 mm/min, S=0.2 mm, t=26 °C. Scanning step does not have a considerable effect on the thickness of sintered layer. The increase in S from 0.1 to 0.2 mm in conditions: P=10 V, V=1000 mm/min, t=26 °C leads to augmentation of sintered layer thickness by 3 %.

References

[1] Polmear I J 1995 *Light alloys: metallurgy of the light metals* (3rd ed. London: Arnold)
[2] Cam G, Kodak M 1998 *Progress in joining of advanced materials* (Int Mater Rev) pp 43–44
[3] Mondolfo L F 1976 *Aluminium alloys: structure and properties* (London: Butterworths)
[4] Bias C T 2009 *Hybrid laser-arc welding* (Woodhead Publishing Ltd: CRC Press) pp. 216–219
[5] Seyda V, Kaufmann N, Emmelmann C 2012 *Investigation of aging processes of Ti–6Al–4V powder material in laser melting. In: Physics Procedia of the 7th international conference & exhibition on photonic technologies LANE* (Fürth, Germany) vol 39 pp 425–431
[6] Gubaidulina R H, Petrushin S I, Galeeva A A 2014 Selecting an Economical Variant of the Manufacturing Method of Engineering Product Fabrication under Current Conditions *Applied Mechanics and Materials* **682** 613-616.
[7] Niu H J, Chang I T 2000 Selective laser sintering of gas atomized M2 high speed steel powder *Mater. Sci.* **35** 31–8.
[8] Asgharzadeh H, Simchi A 2005 Effect of sintering atmosphere and carbon content on the densification and microstructure of laser-sintered M2 high-speed steel powder *Mater. Sci. Eng.*

A **403 1-2** 290
[9] Liu Z H, Zhang D Q, Sing S L, Chua C K, Loh L E 2014 Interfacial characterization of SLM parts in multi-material processing: metallurgical diffusion between 316L stainless steel and C18400 copper alloy *Mater Charact* **94** 116–125
[10] Arkhipov P V, Yanyushkin A S, Lobanov D V 2013 The effect of diamond tool performance capability on the quality of processed surface *Applied Mechanics and Materials* **379** 124–130
[11] Saprykina N A, Saprykin A A, Matrunchik M S 2014 Formation of Surface Layer of Cobalt Chrome Molybdenum Powder Products with Differentiation of Laser Sintering Modes *Applied Mechanics and Materials* **682** 294-298
[12] Belomestnykh V N, Soboleva E G 2012 *Acoustic analogs of elasticity theory ratios for determining anisotropic Poisson ratios of cubic monocrystals* 7th International Forum on Strategic Technology, IFOST – 2012, Tomsk, pp 499-502
[13] Galevsky G V, Rudneva V V, Garbuzova A K, Valuev D V 2015 Titanium carbide: nanotechnology, properties, application *IOP Conf. Ser.: Mater. Sci. Eng.* **91** 012017
[14] Babakova E V, Gradoboev A V, Saprykin A A, Ibragimov E A, Yakovlev V I, Sobachkin A V 2015 Comparison of Activation Technologies Powder ECP-1 for the Synthesis of Products Using SLS *Applied Mechanics and Materials* **756** 220-224
[15] Saprykin A A, Saprykina N A, Dudikhin D V, Emelyanenko S M 2014 Influence of layer-by-layer laser sintering modes on the thickness of sintered layer of cobalt-chromium-molybdenum powder *Advanced Materials Research* **1040** 805-808
[16] Saprykin A A, Petrushin S I, Saprykina N A 2006 The patent of the Russian Federation 2268493
[17] Saprykina N A, Saprykin A A, Borovikov I F, Sharkeev Y P 2015 Influence of layer-by-layer laser sintering conditions on the quality of sintered surface layer of products *IOP Conf. Ser.: Mater. Sci. Eng.* **91** 012031

IOP Conf. Series: Materials Science and Engineering **140** (2016) 012015 doi:10.1088/1757-899X/140/1/012015

X-Ray Diffraction Analysis of the Sintered Y–TZP–Al_2O_3 Ceramics

Nickolai Savchenko[1,2,3,a], Tatiana Sablina[1], Irina Sevostyanova[1], Artem Pshenichnyy[3], Svetlana Buyakova [1,2,3] and Sergei Kulkov[1,2,3]
[1]Institute of Strength Physics and Materials Science SB RAS, 2/4, pr. Akademicheskii, Tomsk, 634055, Russia,
[2] Tomsk Polytechnic University, 30, Lenin Avenue, Tomsk, 634050, Russia,
[3] Tomsk State University, 36, Lenin Avenue, Tomsk, 634050, Russia

[a] Corresponding author: *savnick@ispms.tsc.ru*

Abstract. The paper discusses the wear resistance, friction coefficient, and structure of friction surfaces of fine-grained (0.2 μm) Y–TZP–Al_2O_3 composite rubbed against a steel disk counterface at a pressure of 5 MPa in a range of sliding speeds from 1 to 20 m/sec. It is shown that, starting at 2 m/sec, the fricti on surface is subdivided by a crack network into separate regions within which local spalling occurs at the maximum wear rate and a sliding speed of 5 m/sec. X-ray diffraction reveals inversion (with respect to the initial state) of the peak intensities of the tetragonal phase with random crystalline grain orientation. The degree of this inversion increases with sliding speed. These results are discussed in terms of the effects exerted by the reorientation of martensite-free deformation twins in the tetragonal phase and the formation of a quasi-liquid film on the wear resistance of fine-grained Y–TZP– Al_2O_3.

1.Introduction

Production of various friction pairs using additive technologies - modern trend of Materials. Especially promising is the synthesis of fine-grained (≤0.3 μm) ceramic products. The capability of yttria-stabilized polycrystalline tetragonal zirconia (Y-TZP) ceramics [1] to undergo phase transitions under applied stresses determines their friction behavior [1, 2]. According to [3], when Y-TZP is rubbed against steel at a speed of 1 to 47 m/sec, the dependence of the wear rate on the sliding speed is described by a curve with a peak at 4–5 m/sec. The presence of this peak is attributed to the reduction in the contribution of the transformation toughening mechanism because of the increase in the temperature in the friction area to a level at which the tetragonal phase of zirconia becomes thermodynamically stable. The subsequent decrease in the wear rate at high sliding speeds is caused, according to [3], by the quasi-liquid film on the friction surface of the ceramics, resulting from the melting and oxidation of the steel. To find out the cause of the intensive wear of Y-TZP ceramics, it appears reasonable to study materials with no martensitic transformation and mechanical properties (ultimate strength in bending and fracture toughness) sufficient for catastrophic failure not to occur during high-speed sliding on steel. Such properties are typical of Y-TZP and Y-TZP–Al_2O_3 ceramics [1] with fine-grain tetragonal phase (less than 0.3 μm) being below critical (d_{cr} ~ 1 μm [1]). These ceramics do not undergo martensitic transformation during deformation and fracture [1, 4, 5].
The wear resistance of Y-TZP ceramics with submicron grains has been studied only for relatively low sliding speeds (less than 1 m/sec) [4, 5]. Our goal here is to study the dependence of the wear rate of

Y-TZP–Al_2O_3 ceramics with submicron structure on the sliding speed after dry friction with steel over a wide range of sliding speeds and to study structural changes on the friction surface.

2. Materials and experimental procedure

The samples were prepared from 80 wt.% ZrO_2 (3 mol.% Y_2O_3) + 20 wt.% Al_2O_3 powder produced by decomposing water solutions of Zr, Y, and Al salts in high-frequency plasma. X-ray diffraction revealed three phases in the starting powder: tetragonal ZrO_2, γ-Al_2O_3, and θ-Al_2O_3. After semidry compaction at a pressure of 200 MPa, the samples were nonaxially compressed with a constant true stress of 100 MPa and, in that deformed state, sintered at a temperature of 1300°C for 30 min (for more details on the sample preparation procedure, see [6]). The resulting samples of Y-TZP–Al_2O_3 composite had a density of 94% of the theoretical value, 0.2 μm mean grain size, 600 MPa bending strength, hardness HV = 12.6 GPa, and fracture toughness K_{Ic} = 8 MPa×m$^{1/2}$. After sintering, the samples contained α-Al_2O_3, while the phase composition of zirconia remained the same (tetragonal modification). Tests were conducted on a UMT-1 pin-and-disk friction machine. There was no lubrication, and the sliding speed was increased incrementally. The counterface was a disk made of martensitic cast tool steel (HRC 63–65) with excess $M_{12}C$ carbides and a minor amount of retained austenite. The ceramic samples had rectangular cross-section with a total area of 40 mm^2. The test pressure was 5 MPa, the sliding speed was varied from 1 to 20 m/sec, and the test duration was chosen so that the sliding distance was 2000 m for all speeds. Before each test, the samples were worn in at a speed of 0.1 m/sec and a pressure of 5 MPa. In each test, a computer recorded the friction coefficient at 1 s intervals. The ratio of the material volume lost by the sample during the test to the sliding distance was used as a measure of wear. The ceramic surface before and after friction was examined on a DRON-UM1 X-ray diffractometer with a CuKα-radiation source in the range of angle 2θ from 20 to 80°. Special software was used to do root-mean-square fit of Lorentzian-type profiles to the resulting X-ray patterns. The results were used together with the method described in [7] to determine the size of coherently diffracted domains (CDD). We also analyzed the intensities (I) of X-ray lines (002) and (200) of the tetragonal modification and determined the ratio $I_{(002)}/I_{(200)}$ for each friction mode. The structure of and the element concentration on the friction surfaces were analyzed with a Philips SEM-515 scanning electron microscope and an EDAX ECON IV microanalyzer.

3. Results and discussion

A feature of friction processes in the speed range of interest is release of a great amount of heat, especially at high sliding speeds. The measurement of the contact temperature between the ceramic composite and steel with an infrared pyrometer showed that, at 20 m/sec, the ceramic material heated up to 300°C, while the temperature of the steel did not exceed 200°C; the temperature in the friction areas on the ceramics (flash temperature [8]) was much higher (1600–1800°C at a speed of 10 m/sec [8]). The friction tests demonstrate that as the sliding speed is increased from 1 to 5 m/sec, the wear rate of the ceramics first increases and then (after the speed reaches 20 m/sec) decreases (Figure 1, a). The friction coefficient of the ceramics decreases from 0.8 at low speeds to 0.12 at a speed of 20 m/sec (Figure 1, b). Scanning electron microscopy has revealed grooving (typical of abrasive wear) on the surface after friction at sliding speeds to 2 m/sec. At speeds higher than 2 m/sec, the surface was covered with cracks parallel and perpendicular to the sliding direction, thus forming a network that divides the surface into separate regions. After tests at 5 m/s, the ceramic surfaces display traces of spalling, while after tests at 10 and 20 m/s, the surfaces were as smooth as after the tests at 2 m/sec.

Electron microprobe analysis has revealed that the friction surface of Y-TZP–Al_2O_3 ceramics includes not only elements of the ceramic (Zr, Y, Al), but also elements of the steel counterface (Fe, Mn, W, Ni, Cr, V, Mo). As the sliding speed is increased from 2 to 10 m/sec, the amount of the steel elements increases from 2 to 6 wt.%. Grazing-incidence X-ray diffraction examination of a 1–2 μm surface layer has discovered X-ray peaks of Fe_2O_3. This fact and [3] suggest that a quasi-liquid film was formed on the ceramic surface during friction because of the melting and oxidation of steel.

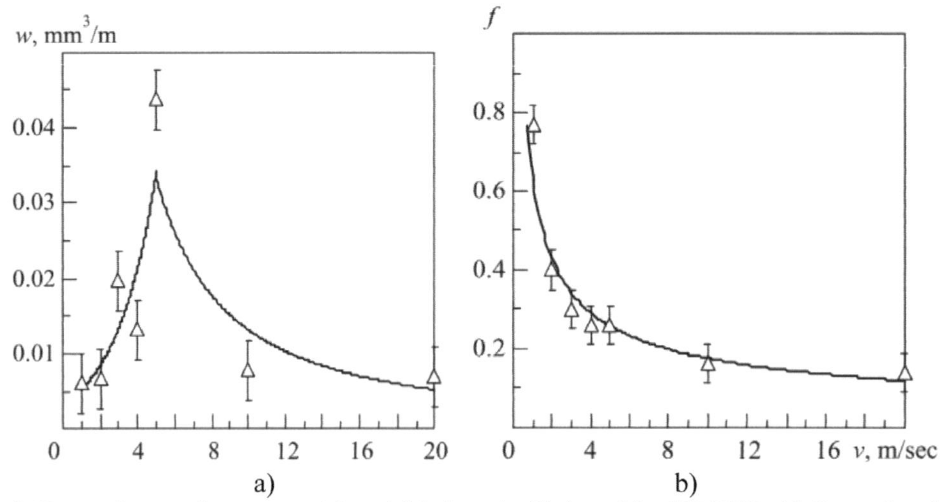

Figure 1. Dependence of wear rate (a) and friction coefficient (b) of Y-TZP–Al_2O_3 on the sliding speed on steel

An X-ray analysis of the Y-TZP–Al_2O_3 surfaces has revealed traces of tetragonal-to-monoclinic martensitic transformation neither on the fracture surface after bending test nor on the friction surfaces. The X-ray texture of the friction surface of the ceramics includes peaks (002) and (200) of tetragonal zirconia whose intensities differ from those on the initial polished surface. After friction test, the ratio $I_{(002)}$ / $I_{(200)}$ exceeds its value (0.8) for the ceramics before the test (Figure 2, a).

It is seen from Figure 2, b that the greater the ratio $I_{(002)}$ / $I_{(200)}$ of the tetragonal phase of ZrO_2, the less the friction coefficient of fine-grained Y-TZP–Al_2O_3 composite. The curve of CDD size of the tetragonal phase on the friction surface versus the sliding speed has a minimum (Figure 3). As the sliding speed is increased to 2 m/sec, the CDD size decreases. With further increase in the speed to 20 m/sec, the CDD size increases, remaining 33% less than in the initial state (Figure 3). According to [11, 12], the CDD size calculated from the (111) peak profile of the tetragonal phase of ZrO_2 can be considered to be the size of intragranular crystallites of the tetragonal phase of ZrO_2.

The results obtained indicate that over the entire range of sliding speeds, the intensity ratio for the doublet (002)–(200) of the tetragonal zirconia (Figure 2) on the surface of the fine-grained Y-TZP–Al_2O_3 ceramics changes so that the c-axis of the tetragonal phase lattice tends to be perpendicular to the friction surface and the average size of its crystallites decreases (Figure 3), especially (to 22 nm) in the region of intensive wear at a sliding speed of 3– 5 m/sec (Figure 1, a). Such structural changes of the friction surface of fine-grained Y-TZP–Al_2O_3 ceramics may be due to high-temperature plastic deformation in the friction area. For example, the observed increase in $I_{(002)}$ / $I_{(200)}$ with increase in the sliding speed (and, hence, increase in the temperature in the friction area) from 1 to 20 m/sec (Figure 2, a) may be caused by the self-consistent motion of coherent twins in grains of the tetragonal phase [9, 10]. Similar relationships between the ratio $I_{(002)}$ / $I_{(200)}$ and the sliding speed were observed after dry grinding of transformation-toughened course-grained (2–3 μm) Y-TZP ceramics [11] and in studying the tribological characteristics in the range of sliding speeds from 0.2 to 11 m/sec [12]. Another possible cause of the increase in $I_{(002)}$ / $I_{(200)}$ may be high temperature plastic deformation leading to a distortion of grains in the sliding direction, which was observed in Y-TZP ceramics after sliding on steel at a speed of 1–5 m/sec [3].

The curve of wear rate versus sliding speed with a peak at medium speeds (Figure 1, a) is similar to that [3] for relatively coarse-grained (2–3 μm) transformation-toughened Y-TZP–Al_2O_3 ceramics of the same chemical composition.

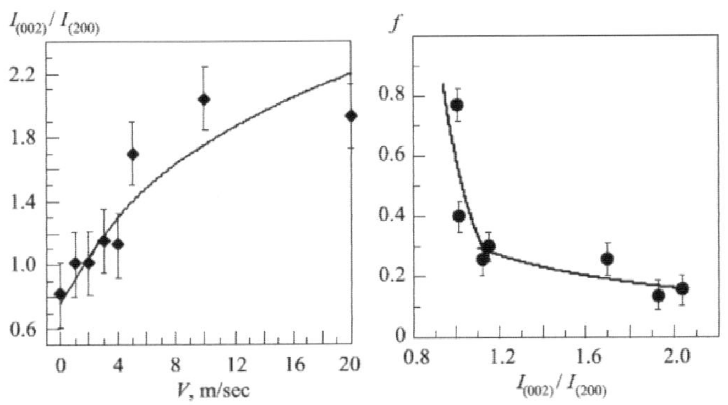

a) b)

Figure 2. Dependence of the ratio $I_{(002)}$ / $I_{(200)}$ on the sliding speed (a) and friction coefficient (b): in Figure 2a, $I_{(002)}$ /$I_{(200)}$ is the ratio of the tetragonal phase on the polished surface of the sample before friction test.

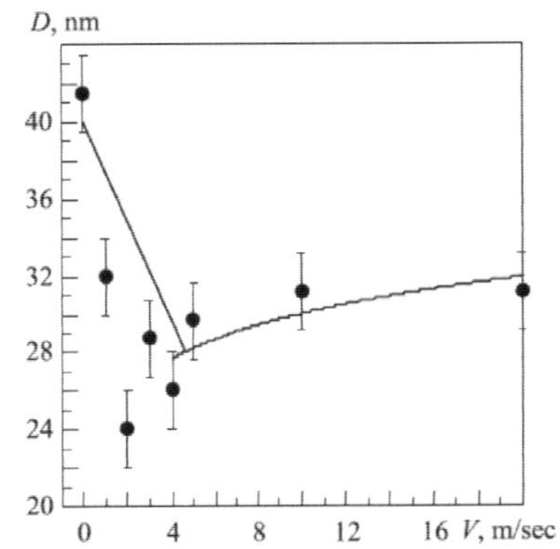

Figure 3. Dependence of the CDD size of the ZrO_2 tetragonal phase of fine-grained Y-TZP–Al_2O_3 composite on the sliding speed.

The presence of the peak indicates that there are at least two factors affecting wear and friction. In the absence of martensitic transformation, the increase in the wear rate with increase in the sliding speed to 5 m/sec (Figure 1, a) may be attributed to the thermomechanical cracking of the friction surface because of the low thermal conductivity of Y-TZP (2 W/(m×K) at 100–1400°C [13] against, for example, 40 W/(m×K) at 25°C for Al_2O_3 [13]), which causes high temperatures in the friction area and high thermal stresses near the surface. Intensive wear is represented by low-scale texture elements on deformed and fractured surface. In such conditions, the friction surface of fine-grained Y-TZP–Al_2O_3 ceramics acquires a microstructure with smallest crystallites of the tetragonal phase (Figure 3) and a crack network with minimum distance between cracks. The quasi-liquid film formed on the friction surface and reducing the contact stresses [3] is responsible for the decrease in the wear rate with increase in the sliding speed from 10 to 20 m/sec. The decrease in the friction coefficient (Figure 2, b) with increasing ratio $I_{(002)}$ / $I_{(200)}$ may be attributed to the intensified plastic deformation of the friction surface, which apparently promotes the formation and fixation of the film on the friction surface. At small (≤2 m/sec) and high (≥10 m/sec) sliding speeds, the wear rate and friction coefficient of the submicron ceramics practically do not differ from those of coarse-grained (1.5 μm) transformation-toughened ceramics of the same chemical composition tested under similar conditions [3]. At medium sliding speeds (~5 m/sec), the fine-grained ceramic composite is worn more intensively (than coarse-grained ceramics [3]), which at high temperatures and plastic strains [3], when the quasi-liquid film has not formed yet, may be due to superplasticity typical of fine-grained ceramic materials.

4. Summary

The dependence of the wear rate on the sliding speed for the fine-grained (0.2 μm) Y-TZP–Al_2O_3 ceramic composite that does not undergo martensitic transformation into a monoclinic phase at a sliding speed of 1–20 m/sec on steel is represented by a curve with a peak at a speed of 5 m/sec. At this speed, the surface is fragmented at different scales: (i) cracks form a network dividing the surface into regions with a minimum average size of 30 μm and (ii) the size of crystallites of the tetragonal phase decreases from 45 nm before friction to 20 nm after friction. It has been shown that the intensity ratio for the doublet (002)–(200) of the tetragonal modification of zirconia is determined by the texture of the friction surface with the c-axis of the crystallite lattice tending to be perpendicular to the friction surface.

Acknowledgement
The work was financed in part by the Ministry of Education and Science of the Russian Federation in the framework of Agreement No.14.575.21.0040 (RFMEFI57514X0040).

References
[1] Nettleship L and Stevens R 1987 *Int. J. High Tech. Ceram.* **3** 1–32
[2] He Y J, Winnubst A J A, Burggraaf A J, et al 1996 *J. Am. Ceram. Soc.* **79** 12 3090–96
[3] Kulkov S and Savchenko 2008 *Epitoanyag* **3** 62 – 64
[4] Kerkwijk B, Winnubst A J A, Verweij H, et al, *Wear* 1999 **225–229** 1293
[5] Basu B, Lee J-H and Kim D-Y 2004 *J. Am. Ceram. Soc.* **87** 9 1771-74
[6] Savchenko N L, Korolev P V, Mel'nikov A G, et al. 2007 *Fund. Probl. Sovr. Materialoved.* **2** 99–101
[7] Rusakov A A, 1977 *X-Ray Diffraction of Metals [in Russian]* (Atomizdat, Moscow) 237
[8] Lim S C and Ashby M F 1987 *Acta Metal.* **35** 1–24
[9] Virkar A V and Matsumoto R L K 1988 Toughening mechanism in tetragonal zirconia polycrystalline (TZP) ceramics *Advances in Ceramics, Science and Technology of Zirconia III* (American Ceramic Society vol 24) ed S Somiya N Yamamoto and H Yanagida (Westerville, OH) pp 653–663
[10] Prettyman K M, Jue J-F, Virkar A V, et al. 1992 *J. Mater. Sci.* **27** 4167–74
[11] Mehta K, Jue J F and Virkar A V 1990 *J. Am. Ceram. Soc.* **73** 1777–79
[12] Korolev P V Savchenko N L and Kul'kov S N 2004 *Pisma ZhTF* **30** 28–34
[13] Samsonov G V, Borisova A L, Zhidkova T G, et al. *1978 Physical and Chemical Properties of Oxides [in Russian]* (Metallurgiya, Moscow) 472

Direct Metal Deposition by Laser in TiNi-Al System for Graded Structure Fabrication

I Shishkovsky[1,2], **N Kakovkina**[1], **F Missemer**[2]
[1] P.N. Lebedev Physical Institute (LPI) of the Russian Academy of Sciences, Samara branch, 443011 Samara, Russia
[2] Ecole Nationale d'Ingénieurs de Saint-Etienne, 58 rue Jean Parot,42023 Saint-Etienne, France

E-mail: shiv@fian.smr.ru

Abstract. Intermetallic phase formation was studied in powdered TiNi-Al system under layerwise laser cladding with the aim of forming a gradient of properties due to a change in the concentration relation of Al in the NiTi powder mixture from one layer to another. The relationship between the laser cladding parameters and the intermetallic phase structures in consecutively cladded layers were determined. The structure of intermetallic compounds formed by laser synthesis was studied by optical microscopy, measurement of microhardness, SEM with EDX analysis. Microhardness doubling from 500 HV to 1000 HV was achieved due to nitinol matrix enrichment by Al, which is promising for aerospace applications.

1. Introduction

The application range of nickel superalloys is diverse and covers gas turbines for electro- or gas-pumping stations, rocket engines, automatic spacecraft and nuclear reactors. The basic units of turbines where nickel-based super alloys could be used are combustion chambers, guide blades in nozzles, rotor blades or turbine disks. TiAl-based intermetallics also have a number of advantages over the conventional titanium alloys, such as a higher modulus of elasticity, lower density, better mechanical properties at high temperatures. The study of peculiarities of phase structure transformations of TiNi-Al alloys (including those doped with aluminum), which can affect mechanical properties of products made by additive technologies using these powders, is an urgent problem. Due to such complex intermetallic phases, they could replace refractory nickel superalloys in future.

There are five mechanisms for strengthening superalloys: solid solution, dispersion, grain-boundary, deformation and textural strengthening [1]. The first three ones depend on the alloy nature. In nickel-based alloys the strengthening occurs essentially by the dispersive mechanism due to the release of γ–$Ni_3(Al,Ti)$ phase. In our paper [2], laser beam (LB)-aided control of self-propagating high-temperature synthesis (SHS) in Ni-Al, Ni-Ti, Ti-Al systems for layerwise manufacturing of three-dimensional (3D) parts was suggested and experimentally realized for the first time.

Functionally Graded Structures (FGS) and FG objects fabricated by applying complex and dissimilar materials ensure specific properties of final product. The manufacturing of 3D FG objects by 3D laser

[1] Shishkovsky I, e-mail: shiv@fian.smr.ru

cladding (LC) and/or Direct Metal Deposition (DMD) is one of the most promising techniques capable of meeting various industrial challenges [3, 4]. The possibility of controlling the hardness of graded structures from layer to layer by changing powder composition and introducing reinforcing intermetallic inclusions into superalloy matrix widens the range of possible applications of 3D parts in aerospace and nuclear industries.

In the present study, it is offered to perform FGS fabrication in a TiNi-Al system by layerwise increase of Al concentration from the substrate up to the surface during 3D LC. Due to very different melting temperatures in high-Al (i.e. rich in Al) and low-Al regions, the identification of phase structures in the whole compositional range of TiNi-Al up to 1000 °C would be a very interesting task [4-7]. Optical microscopy (OM) and scanning electron microscopy (SEM), X-ray phase analysis and microhardness measurement were employed in this study in order to analyze the FGS.

2. Experimental procedure

The following materials were used for the FGS fabrication: aluminum powder with 99 wt% of Al and nitinol powder both produced by TLS Technik GmbH & Co (Germany) containing 99.76 wt% of intermetallic NiTi phase. The NiTi powder was represented by a 60 µm fraction. The pre-alloyed NiTi phase was obtained by means of gas phase atomization. The NiTi stoichiometric ratio was 45 wt% of Ti and 55 wt% of Ni. The Al powder particles were mainly spherical with the size of ~60-80 µm for 95% of them.

The substrates made of Ti-6Al-4V were round plates with the diameter of 65mm and the height of 5 mm. The powder size distribution was studied by means of a granulomorphometer ALPAGA 500NANO (OCCHIO s.a.).

All the experiments were carried out in the DIPI laboratory (ENISE, France) with the aid of the HAAS 2006D (Nd^{+3}:YAG, 4000 W, cw) with laser beam delivery system, powder feeding system, coaxial nozzle, and digitally controlled 5-axes table [4]. Some features of the equipment are reported below:

- Powder feeder: 2-channel MEDICOAT, the powder feeding rate can be adjusted separately for each channel. Argon and nitrogen are used as carrying gases.
- Coaxial nozzle: the advantages of the coaxial injection are small heat affected zone (HAZ) and possibility of multidirectional cladding due to radial symmetry between the laser beam and powder flux. The shielding gases (Ar, N_2) are surrounding, focusing and protecting the powder flow and the melting pool from oxidation or, vice versa, causing the nitridation initiation.
- CNC center LASMA 1054 is applied for the displacement of the sample and nozzle relative to each other with the positioning accuracy up to 1 µm.

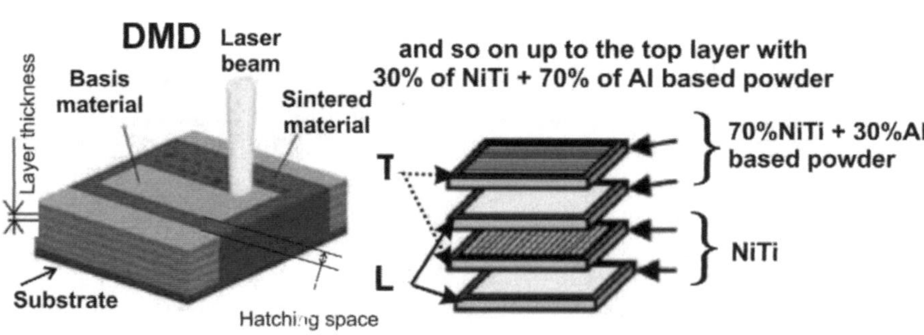

Figure 1. Schematic of the multi component graded structure fabrication by 3D laser cladding. Longitudinal – L; transversal –T.

The method of the functional graded structures (FGS) fabrication used in the present study is schematically presented in Figure 1, as it was proposed earlier [4]. The hatching distance was 2 mm, the layer depth was ~2 mm and the powder feeding rate was ~10 g/min. The layers were made of NiTi and Al powders on a steel substrate with the following strategy: the first two layers were of pure NiTi, the two next consisted of 70% NiTi + 30% Al-powder, the third couple of two equal layers – 50% NiTi + 50% Al-powder and at last, the two upper layers, 7th and 8th, had the ratio of 30% of NiTi + 70% of Al-powder. Each second layer was formed on the bottom layer after turning it by 90 degrees. Argon was the carrying gas. The laser scanning speed V_L was 400-600 mm/min, laser power P varied within the range of 800–1200 W and laser beam (LB) spot diameter d_L was 3 mm. The first channel of the feeder with NiTi powder had the gas flow rate of approximately 20 L/min, while the second one with Al powder, ~10 L/min.

The macrostructure of 3D samples was examined with the Neophot 30M optical microscope with digital camera and PMT-3M (OKB SPECTR Ltd., St. Petersburg, Russia) microhardness tester. The morphology of the laser cladded layers was studied with the use of LEO-1450 scanning electron microscope (Carl Zeiss Microscopy GmbH, Jena, Germany) equipped with an energy-dispersive x-ray analyzer (INCA Energy 300, Oxford Instruments Analytical, High Wycombe, UK).

3. Results and discussion

In our opinion, the choice of metal used as the matrix material, the type and proportion of the component added to synthesize the intermetallic phase are fundamentally important for LC in Ni-Ti-Al system. Since laser irradiation (LI) is well reflected by aluminum, the more refractory nitinol could be used as the matrix material, and in accordance with the Ni-Ti-Al phase diagram, it is reasonable to expect the emergence of new stable intermetallide phases of aluminides, including γ $Ni_x(Al,Ti)$ phases. The optical metallography (OM) is presented in Figure 2. The bottom cladded layers are significantly different in terms of structure and chemical composition as compared to the top layers. The bottom layers have dendrite structure with clearly visible laser passages. Some porosity can be noticed in the sample volume (Figure 2, bottom). The top layers are the mixture of globular inclusions of different nature. Initial grey nitinol matrix was locally preserved. At last, the middle layers have a complex intermetallic structure, which we expected to use for accounting real strengthening due to the formation of the $Ni_x(Al,Ti)$ phases. Boundaries between layers are clearly visible, especially in place of Al content change.

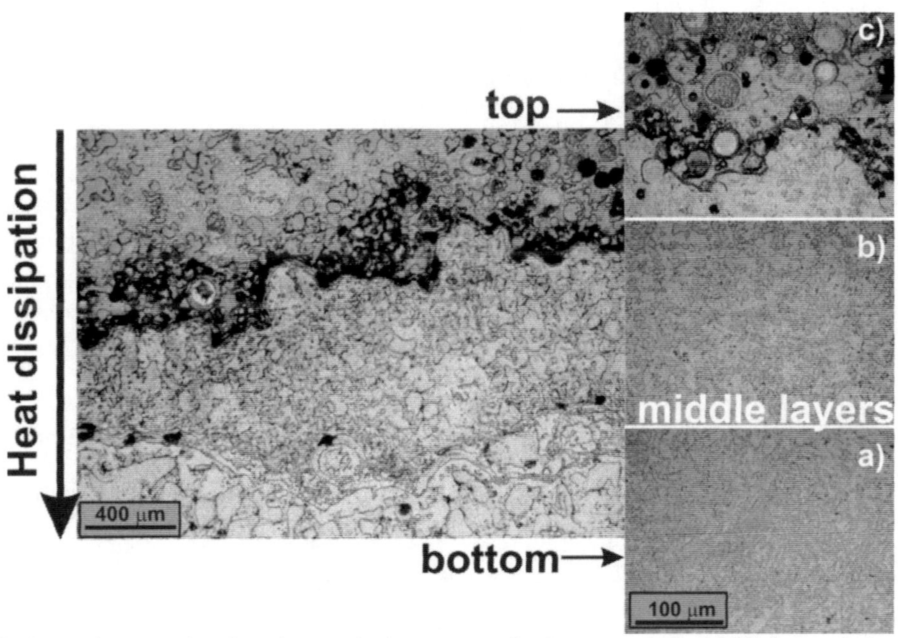

Figure 2. OM micrographs showing typical macro and microstructures of 3D laser clad coating of NiTi-Al multilayer system

The results of microhardness testing are shown in Figure 3. In the vicinity of the substrate, the microhardness is equal to 500-550 $HV_{0.1}$. With increasing distance from the substrate, the hardness increases and reaches 800-1000 $HV_{0.1}$. We connect the separate dips in microhardness in the bottom layers (up to 2 mm from the substrate, Fig. 3) with excessive porosity, and in the top layers, with the indenter entry into the complex $Ni_x(Al,Ti)$ intermetallic phases. On the whole, the measured microhardness values significantly exceed other researchers' data (500-550 $HV_{0.1}$ for the SLM of the NiTi by [8]) and could be estimated as a promising result of our study.

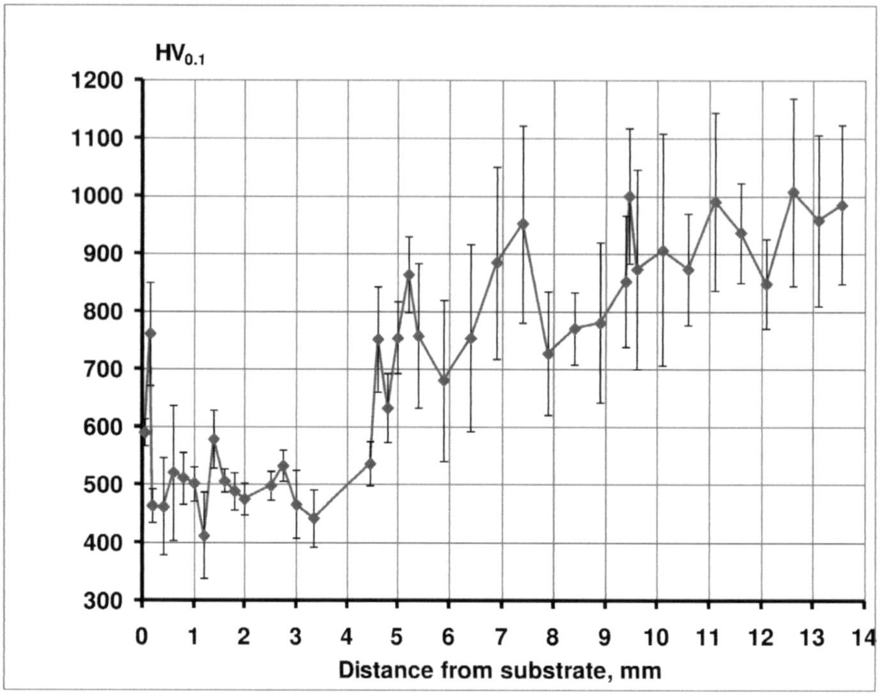

Figure 3. Microhardness of NiTi-Al FG multilayer system. Influence of distance from the substrate.

The SEM images of the micro- and substructures are shown in Figure 4 and enable more thorough study of microstructures appearing during the FG multilayer fabrication. The SEM images correspond to the bottom, middle and top layers of the FGS after the layerwise LC in the Ni-Ti-Al system. The top series in Figure 4 represent the SEM images of substructures under higher magnification than those in bottom series of the figure. It is evident that these substructures (Figures 4a and 4b) have dendrite nature.

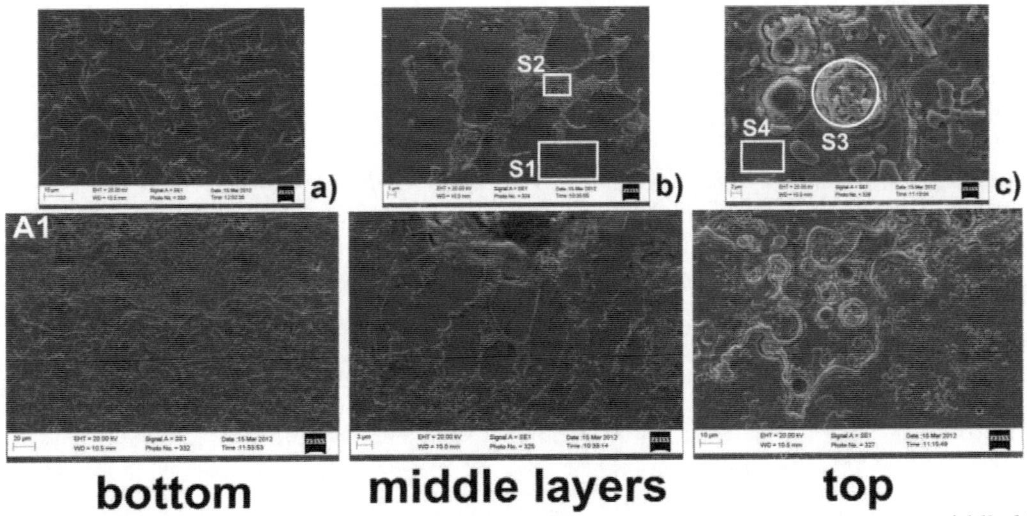

Figure 4. SEM micrographs showing typical solidification microstructures of bottom a), middle b), and top c) clad coatings with EDX results of NiTi-Al multilayer system.

The elemental analysis of regions S1-S2, S4 and A1 and the centesimal analysis of Ni-Ti–Al content also testifiy that we deal with the formation of a complex intermetallic phase. Ti41.22Ni38Al20.78, Ti27.37Ni46.63Al26 and Ti39.99Ni35.63Al24.38 complex intermetallic phases, prepared by laser layerwise cladding, mainly consist of TiNi intermetallic compound exhibiting high hardness. Jung et al [9] considered that the precipitation of Ni_2TiAl Heusler phase with $L2_1$-structure in a supersaturated *B*2-TiNi matrix and significant strengthening could be observed for NiTi-Al system under high temperatures.

Table 1. EDX data by Figure 4

Element, wt%	Ni K	Ti K	Al K	O K	Total
S4 area	38.00	41.22	20.78	--	100.0
S3 circle	36.22	44.23	9.37	10.17	100.0
S2 area	46.63	27.37	26.00	--	100.0
S1 area	35.63	39.99	24.38	--	100.0
A1 all area	46.43	53.57	--	--	100.0

4. Conclusion

The applicability of 3D laser cladding for creating a functional gradient and building of $Ti_xNi_yAl_z$ intermetallic structures was experimentally studied. It was shown that under layerwise LC, the formation of a $Ni_3(Al,Ti)$ intermetallic compound is observed in a NiTi-Al system.

Two-fold increase of microhardness from 500 HV to 1000 HV was obtained for similar samples due to the change of the element relation in the NiTi-Al system after 3D laser cladding.

Hence, the possibility of controlling the hardness of a multi-layer structure by changing powder composition was shown. Appropriate aluminum doping into nickel titanium can expand the range of 3D FGM application in aerospace or nuclear industries.

Acknowledgments

This study was supported by the Russian Foundation for Basic Research (grant No. 14-29-10193 ofi-m).

References

[1]. Canh R and Haasen P 1983 *Physical Metallurgy* 3rd revised and enlarged edition (Amsterdam, Oxford, New York, Tokyo: North-Holland physics publishing) 1983 Part 1 - 2 and 3 p 2050
[2] Shishkovsky I, Makarenko A and Petrov A 1999 *Combustion Explosion and Shock Waves* **35**(2) 166
[3] Ocylok S, Weisheit A and Kelbassa I 2010 *Physics Procedia* **5** 359
[4] Shishkovsky I, Missemer F and Smurov I 2012 *Physics Procedia* **39** 382
[5] López F and Alvarez R 2012 Development of Sintered MCrAlY Alloys for Aeronautical Applications *Sintering - Methods and Products* ed Shatokha V (Rijeka: InTech Publ) pp. 47-72
[6] Chen Y and Wang H 2003 *Materials Letters* **57** 2029
[7] Wang C, Li T, Yao B, Wang R and Dong C 2010 *Surface & Coatings Technology* **205** 189
[8] Shishkovsky I, Yadroitsev I and Smurov I 2012 *Physics Procedia* **39** 447
[9] Jung J, Ghosh G, Isheim D and Olson G 2003 *Metallurgical and Materials Transactions A* **34**(6) 1221

Mechanical Properties of Zirconium Ceramics with Hierarchical Porous Structure

S Kulkov[1,2,3], E Shutilova[1], S Buyakova[1,2,3]
[1]Institute of Strength Physics and Materials Science SB RAS, 2/4, pr. Akademicheskii, Tomsk, 634055, Russia
[2] Tomsk Polytechnic University, 30, Lenin Avenue, Tomsk, 634050, Russia.
[3] Tomsk State University, 36, Lenin Avenue, Tomsk, 634050, Russia

Email: kulkov@ms.tsc.ru

Abstract. The work studies porous ceramics produced from ultra-fine powders. The porosity of ceramic samples was from 15 to 80%. The ceramic materials had cellular structure. A distinctive feature of all deformation diagrams obtained in the experiment was their nonlinearity at low deformations, which was described by the parabolic law. It was shown that the observed nonlinear elasticity for low deformations on deformation diagrams is due to mechanical instability of cellular elements in a ceramic frame.

1. Introduction

Today, the use of additive technologies in the production of structural and functional application of ceramics is a promising direction. Additive technologies allow producing objects of complex shape.

Ceramics based on partially stabilized zirconia are the most interesting among the variety of ceramic materials due to their inherent high fracture toughness because of transformation toughening phenomena. Their characteristics are known to be determined by the quality of source ceramic powder (particle shape, particle size distribution), the conditions of compaction and sintering modes and any features of each phase, and how these phases, including pores, are arranged in relation to each other. The most important factor in successful application of materials is understanding the features of a structure emerging in them and their behavior under mechanical impact.

Plasma spray synthesis and chemical co-precipitation methods are the main efficient routes of ultra-fine powder production for additive processes to obtain ceramics due to activation of sintering [1]. The sintering process for these powders with identical chemical composition may be very different, and the final structure of a sintered body depends on particle size, surface energy, strain conserved in the whole system, etc. [2]. For example, one can obtain hollow-ball particles, which shape will condition a special morphology structure of materials [3, 4].

The aim of the work is the investigation of densification, structure and mechanical properties of materials from zirconia-based powders produced by plasma spray synthesis and sintered at different temperatures.

2. Materials and experimental procedure

Ceramic samples made from plasma-sprayed and chemically precipitated (97 mol% ZrO_2 + 3 mol% Y_2O_3) powders were studied. To study densification kinetics, the samples were produced by pressing as-received powders till relative density of 0.33 and sintering at 1500 °C for up to 20 hours.

Structure investigation was carried out for the samples sintered in the temperature range 1400–1650 °C during isothermal holding for 1 to 5 hours after uniaxial compression with the loading speed of $4*10^{-4}$ s^{-1}.

Phase identification and evaluation of coherently diffracted domains (CDD) were determined from X-ray diffractometry data [5, 6]. Scanning electron microscope observations operated at 20-30 kV were used to determine the structure and average grain and pore size.

3. Results and discussion

Zirconia powder was characterized by spherical particles and their agglomerates (Figure 1a). The average particle size was 1.5 µm. The specific surface of chemically precipitated powder was measured to be 7 m^2/g. According to the X-ray data, the tetragonal phases of ZrO_2 were predominant in the amount of 95% with an average CDD size of 20 nm. An average CDD size of monoclinic phase was equal to 20 nm.

Density dependencies during sintering process are represented in Figure 1b, and it can be concluded that most intensive densification occurred at the heating stage. The analysis of this dependence using equation $\Delta L/L = K \cdot \tau^n$ ($\Delta L/L$ is relative shrinkage, K is kinetic coefficient; n is constant of densification rate) in log-log coordinates revealed that *n* for the samples made from plasma-sprayed powder is twice as much as for the samples based on chemically precipitated powder: 0.1 and 0.04, respectively [7].

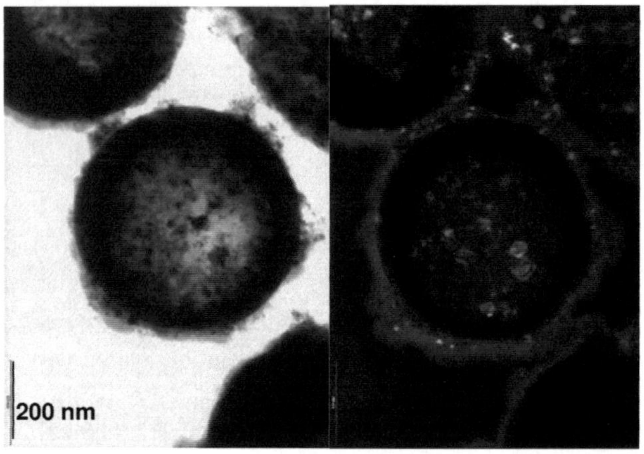

a

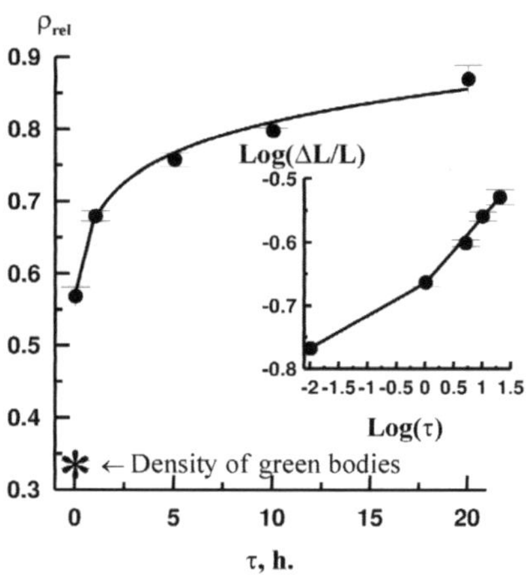

b

Figure 1. ZrO_2 powder synthesized by plasma-sprayed method: a) TEM image, bright and dark filed; b) dependencies of relative density on the duration of isothermal holding for ZrO_2 powder and kinetic dependences of samples shrinkage during isothermal holding.

X-ray analysis has shown that the tetragonal phase content in sintered ceramics was decreasing with the increase of the holding time up to 5 hours for materials based on plasma-sprayed powder from 95% down to 60%; further increase of holding time had no effect on the phase composition.

The structure of ceramic materials produced from plasma-sprayed ZrO_2 powder was represented as a system of cell and rod structure elements (Figure 2a). Cellular structure formed by stacking hollow-powder particles can be easily seen on the images of fracture surfaces of obtained ceramics. There were three types of pores in ceramics: large cellular hollow spaces, small interparticle pores which are not filled with powder particles and the smallest pores in the shells of cells. The cells generally had irregular shape. The dimensions of the cells are far larger than the thickness of walls, which are presented by single-layer ZrO_2 grains.

The increase of the pore space in the ceramics was accompanied by the decrease of the average grain size. Quantity and the size of pores and the grain size in the materials produced by powder technology are highly dependent on thermokinetic sintering conditions (Figure 2b). Obviously, the increase in the volume of pores in the material from ~30% up to 80% was achieved by reducing the sintering temperature of the samples, and it was accompanied by an increase in the average size of large pores from 2 to 6 microns. Changing the porosity of the material had practically no influence on the average size of interparticle pores, the average size of which was 0.5 microns. It can be assumed that the presence of large pores close to a spherical shape in the tested ceramics is due to the presence of hollow spherical particles in source powders, since their average size is commensurate with the average size of large pores in the sintered material.

a

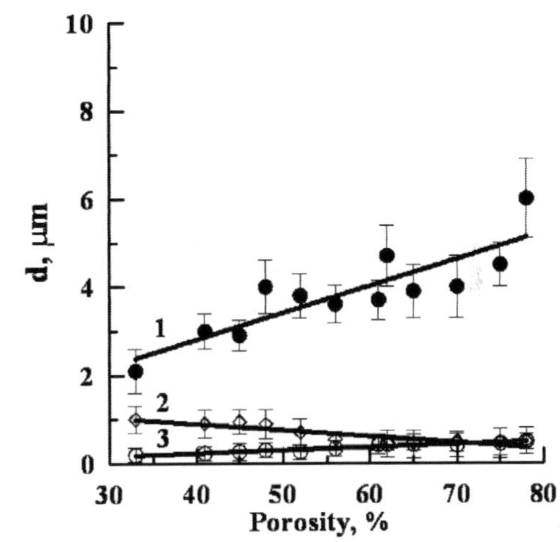

b

Figure 2. a) Fracture surface of sintered ceramic samples. Cellular structure of porous ceramics (porosity is 30%); b) The dependence of average size of the interior of the cells (1), of average size of grain (2) and of average size of small interparticle pores (3) on porosity.

Stress-strain diagram of porous ceramics, which were produced by plasma-sprayed method are presented in Figure 3a. The obtained stress-strain diagrams had descending branch with a monotonic decrease of stress. This is the evidence of damage accumulation in the samples in contrast to the stress-strain diagrams of brittle materials with a homogeneous structure. Microdamages appearing in the material has local nature and the sample under load retained the ability to resist increasing load. A distinctive feature of all the σ - ε diagrams obtained in the experiment was their nonlinearity at low deformations which was described by the parabolic law. Cyclic loading of samples on the parabolic section of the diagrams did not reveal residual strain. Therefore, the nonlinearity in the stress-strain diagrams was due to the elastic deformation of ceramics with cellular structure.

Replotting deformation diagrams in ln-ln coordinates allowed us to determine the exponent of the Hollomon equation [8] ($\sigma = K\varepsilon^n$, where σ is true stress; ε is true strain; n is parabolic index; K is constant for a given material, defined as a value of true stress for small true strain) from the experimental data. In this case, the index takes the value of the power function of the slope of the strain diagram in logarithmic scale (Figure 3b).

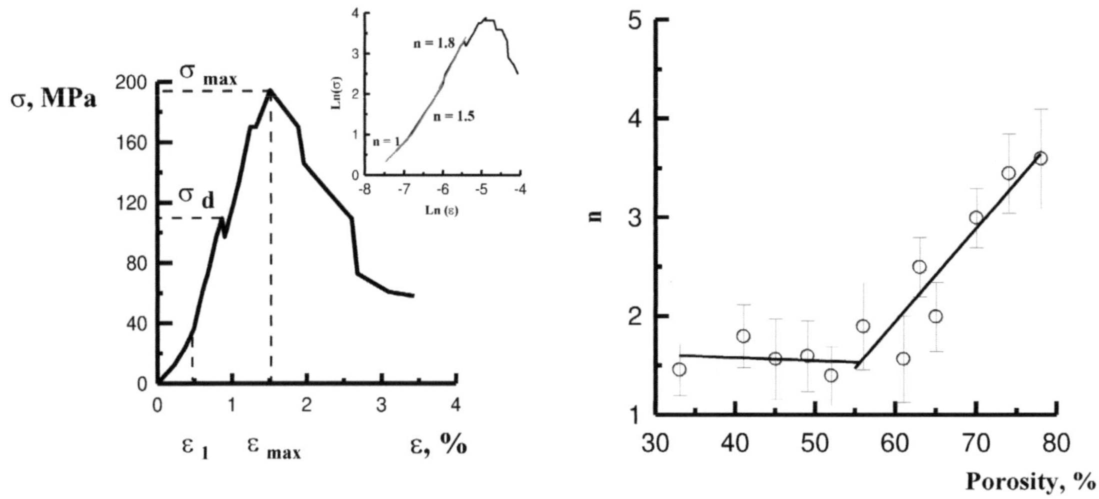

a *b*

Figure 3. a) Stress-strain diagrams of ceramics compression with the porosity of 50%; b) the dependence of parabolic index from porosity.

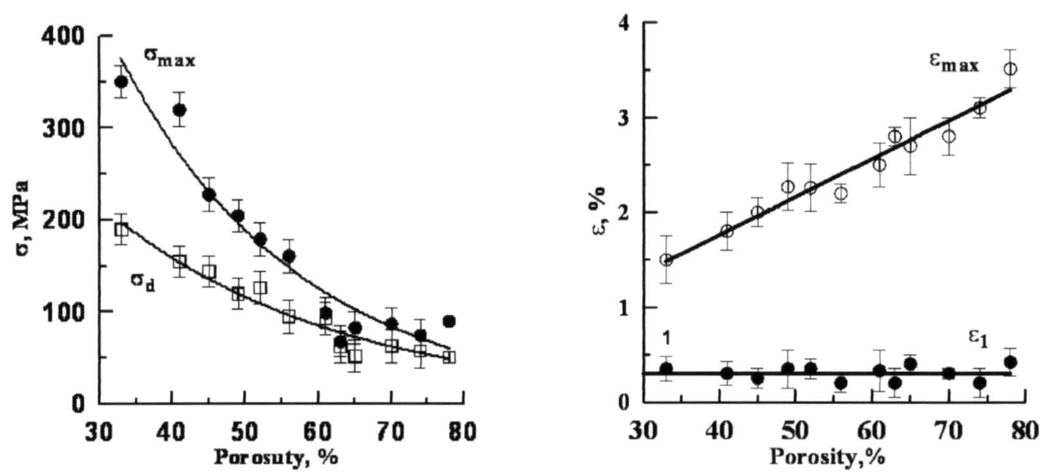

a *b*

Figure 4. a) The dependence of maximum stress σ_{max} and the stress corresponding to emergence of microdamages in material σ_d from porosity; b) The dependence of strain corresponding to the end of the parabolic section on the rising branches of the stress-strain diagrams ε_1 and the strain corresponding to the maximum value of stress in ceramic samples ε_{max} from porosity.

The dependencies of maximum stress σ_{max} and the stress corresponding to the emergence of microdamages in material σ_d from porosity are presented in Figure 4a. Evidently, while the porosity decreases in the ceramics from 80% down to 30%, the maximum stress and the stress corresponding to the emergence of microdamages in material increased from 50 to 400 MPa and from 50 to 250 MPa, respectively. With the increase in the quantity of pores in the ceramics, the difference between the maximum stress and the stress corresponding to the emergence of microdamages decreased. When the volume fraction of pores in the samples was more than 60%, σ_{max} and σ_d were identical. Rising and descending branches of stress-strain diagrams became gentler.

The dependence of strain corresponding to the end of the parabolic section on the rising branches of the stress-strain diagrams ε_l (1) and the strain corresponding to the maximum value of stress in ceramic samples ε_{max} (2) from porosity are presented in Figure 4b. The increase of porosity from 30% up to 80% caused the increase of ε_{max} from 1.5% to 3.5%. The obtained values of relative strain (up to ~3.5%) were significantly higher than values for nonporous ceramics. The increase of the porosity in ceramics had virtually no impact on the amount of strain ε_l, which averaged to 0.5%.

4. Conclusion

It has been shown that the most intensive densification of studied materials took place during heating stage. In the stress-strain diagrams, the nonlinearity occurred due to the elastic deformation of ceramics with cellular structure. The character of the received strain-porosity dependences probably was a result of porosity type change.

It was shown that the "stress - strain" diagrams on the initial stage of deformation was nonlinear with high parabolic factor of strain-stress curves. It was shown that rod-like and/or cellular structures were formed in material, and the fracture of material occurred in the elastic area.

Acknowledgement

This work was financially supported by Tomsk State University Competitiveness Improvement Program and Siberian Branch Program #III.23.2.3.

References

[1] Fedorchenko I M and Ivanova I I 1974 *About influence of particle size and specific surface on densification during the sintering* (Kiev: Naukova Dumka)

[2] Gusev A I and Rempel A I 2001 *Nanocrystalline materials* (Moscow: PHYSMATHLIT)

[3] Kulkov S N and Buyakova S P 2007 Structure, phase composition and technologies features of zirconia-based nanosystems *Russian Nanotechnology* **2 1** 119-132

[4] Kalatur E, Buyakova S P and Kulkov S N 2014 Porosity and mechanical properties of zirconium ceramics *Epitoanyag - Journal of Silicate Based and Composite Materials* **66 2** 31-34.

[5] Torayda H, Yoshimura M and Somiya S 1984 *J. Am. Ceram. Soc.* **67** 119-121

[6] Gorelic S S, Skakov Yu A and Rastorguev L N 1994 *X-ray and electron/optical microscope analysis* (Moscow: MISIS)

[7] Kulkov S N, Buyakova S P and Gomze L A 2014 Structure and mechanical properties of ZrO_2 – based systems *Epitoanyag - Journal of Silicate Based and Composite Materials* **66 1** 2-6.

[8] Hertzberg R W 1989 *Deformation and fracture mechanics of engineering materials* (John Wiley and Sons, Inc.).

IOP Conf. Series: Materials Science and Engineering **140** (2016) 012018 doi:10.1088/1757-899X/140/1/012018

Coating synthesis controlled by electron-beam heating

A I Gordienko[2], A G Knyazeva[1,31], I L Pobol[2]
[1] Tomsk Polytechnic University, Tomsk, Russia
[2] Physical-Technical Institute of the National Academy of Sciences of Belarus, Minsk, Republic of Belarus
[3] Institute of Strength Physics and Material Science of the Siberian Branch of the Russian Academy of Science, Tomsk, Russia

E-mail: anna-knyazeva@mail.ru

Abstract. The methods of combined electron-beam treatment of parts made of steel with one- and two-layer coatings are studied experimentally. Ti-Ni, Ni-Al and Al-Ti systems were used as the examples in the experiments. The mathematical model is suggested for coating formation in the controlled regime of high temperature synthesis during high energy source motion along the preliminarily deposited layer of exothermic composition. The study takes into account the difference in thermophysical properties of the materials of coating and substrate, heat release from chemical reaction that leads to the coating properties formation and other factors. The realization of the synthesis depends on technological parameters. Various regimes of the treatment process are investigated numerically.

1. Introduction

Self-propagating high-temperature synthesis (SHS) methods possess the extensive possibilities to create the materials of various types [1-3]. Coating synthesis on metals using SHS faces some difficulties and is not a widely spread technological process. A material obtained by SHS, as a rule, has the porosity comparable with initial one and has low mechanical characteristics. To obtain compact and chemically uniform material, the following treatment is necessary.

Employment of electron-beam has advantages for a broad spectrum of technology applications [4-6], including surface engineering: thermal treatment of alloys based on iron, titanium, aluminum and copper in the hardening regime from solid and liquid states, build-up welding of coatings on these alloys from materials with considerably different properties.

It is to be expected that a combined method of electron-beam treatment of material with preliminarily deposited powder alloy when the coating synthesis is realized allows for realizing the advantage of electron-beam treatment and condensed phase synthesis if the technological process will be arranged correctly for obtaining a material with anticipated properties. Preliminary investigations with green compositions of powders and composite wires of Ti-Ni, Ni-Al, Al-Ti [4] have shown that electron-beam technologies enable obtaining the smooth transition of the distribution of composition, chemical elements and mechanical properties from deposited layer to the basic detail. Intermetallic coating synthesis during laser heating was described in [7]. Here, the model was suggested taking into account the reaction retardation and controlling of the heating by moving source. However, only in [8] the heat losses from reaction zone to the substrate were taken into account, which could be the key

[1] anna-knyazeva@mail.ru

factor for the realization of one or another synthesis regime. Some general consideration in conjugate and coupling models formulation were recounted in [9]. But only certain results for the simplest variant of the model (without reaction retardation by product layer) were presented in the previous work.

Note, that on the one hand, Ti-Ni, Ni-Al and Al-Ti systems, especially in the form of coatings on the base of steels and titanium alloys, are weakly examined despite enormous number of publications, and on the other hand, they are extremely promising from the practical point of view. Really, they could be applied to heighten the wear resistance and corrosion stability of details made from inexpensive engineering materials. For example, Ni-Al coating can be used for anticorrosion insulation of large vessels, reservoirs, metallic barriers and etc. γ-TiAl is the most perspective composite for automotive (in turbo compressors) and for aerospace applications [10-15]. This material has low density, high heat resistance and high-temperature strength. The production of blading for aviation gas-turbine engines from this material is a pretty complex problem. One of Ti-Al alloy applications can be their deposition on a titanium-based alloy using the combined method EB+SHS.

This paper briefly describes the experimental procedure for coating synthesis from Al-Ni (this system is actively studied in the applications of SLS and SLM [16, 17], and during development of new alloys [18]), shows the results of investigation of phase composition determination after electron-beam treatment and suggests the model of phase composition evolution for the coating synthesized on the substrate.

2. Experimental results

During experiments, Ni-Al coatings were deposited on the surface of the part by thermal spraying. The material in the form of a wire was sprayed using EM 12M setup and deposited on the substrate prepared from steel (0.2%C, 0.5%Mn). On sputtering, Ni and Al react with each other to form intermetallic compounds. The 0.1 mm-thick coatings consisting of one (Al or Al–Ni) or two (Al + Al–Ni) layers have lamellar structure; they demonstrate defective interface between the substrate and sublayer; contain unfused particles and have substantial porosity. To improve the cohesion strength and diminish the porosity, the EB energy has been used. Coating structure ordering occurs during intensive heating and fusing. The formed liquid phase enables the acceleration of surface and volume diffusion; gas desorption occurs, and the diffusion zone forms between the coating and the base. The formation of dendritic structure oriented mainly in perpendicular to the surface is typical for the coatings. This is conditioned by the heat sink direction during the melt solidification when EB action is terminated. The structure becomes more dispersed near the substrate. The diffusion zone is observed between the aluminum coating and the base.

The coating phase structure changes during EB treatment. The concentration of NiAl, Ni_2Al_3 and Al_2O_3 is increased, new phases of Ni_2Al, Fe_3Al and FeAl are formed and the traces of NiO and Ni are observed. The content of hard phases (NiAl, Al_2O_3, NiO) increased under heating by the source at relatively low intensity (energy density was $0.8 \cdot 10^3$ W/cm^2). When the heating is more intensive (specific power is $2 \cdot 10^3$ W/cm^2), the partial interaction between coating and substrate materials occurs, and the portion of iron-containing Fe_3Al, FeAl phases and plastic phase Ni_2Al_3 is increased. This leads to the change in coating properties depending on the density of heating source energy. By selecting the proportions of Ni and Al, one can control various properties of materials produced due to exothermic reaction.

As a result of surface coating reflow by electron beam, the coating-substrate adhesion strength is increased by the factor of 15 to 20 reaching 140 MPa, while the wear resistance of the coating-substrate system is enhanced 2.5 times as compared with initial state.

The corrosion resistance of samples increased 3-6 times in 3% NaCl and 5% H_2SO_4 solutions. When modified coatings were subjected to blasting with SiO_2 particles at the rate of 40 m/s, their wear resistance doubled versus as-deposited state. The modified Ni-Al coatings can be efficiently used for protection of surfaces under the conditions of their simultaneous exposure to high temperatures and erosive media.

3. Mathematical model

A complete thermophysical model of controlled synthesis of the coating in Cartesian coordinate system includes three-dimensional thermal conductivity equation for the base (subscript b) and thermal conductivity equation for the preliminary deposited layer (subscript s), where the physical chemical conversions can proceed.

$$c_k \rho_k \frac{\partial T_k}{\partial t} = -\nabla \cdot \mathbf{J}_k + W_k + \Phi_k ,$$

where $k = b, s$; c_k, ρ_k are heat capacities and densities of materials; heat fluxes J_k obey to Fourier law; W_k, Φ_k are total sources of heat due to chemical conversions and external heating. If external source acts only in the coating, then $\Phi_b = 0$.

The axis arrangement is shown in Figure 1.

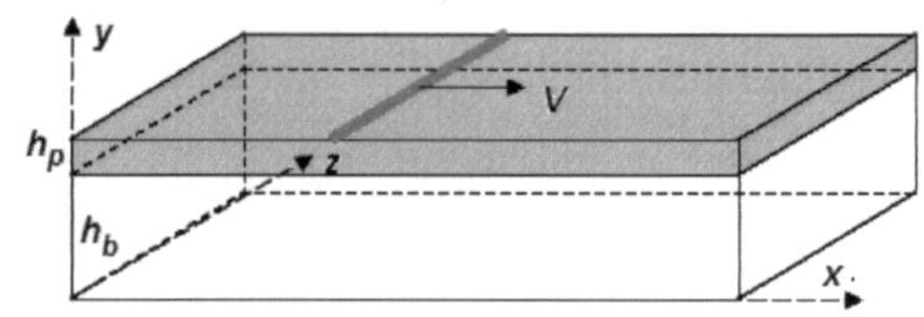

Figure 1. Problem formulation

Under the condition of volume absorption of the energy of moving source developed to the line, the energy source for the coating can be written as

$$\Phi(x, y, z, t) = q_0 \, exp\left[-\frac{(x - Vt)^2}{{R_0}^2}\right] \varphi_S(y, z),$$

where R_0 is electron beam radius, V is its motion velocity along the Ox -axis.

The condition of ideal thermal contact is correct in the coating-base interface

$$\lambda_b \left(\frac{\partial T_b}{\partial y}\right)_{y=h_b-0} = \lambda_p \left(\frac{\partial T_p}{\partial y}\right)_{y=h_b+0} ; \; T_{b\,y=h_b-0} = T_{p\,y=h_b-0} .$$

The heat losses by irradiation (according to Stephen-Bolzman law) are possible in the surfaces $y = 0$ and $y = h_b + h_p$. Other surfaces are insulated.

Chemical reactions can occur in the coating and lead to phase composition change. In this case,

$$W_p = -\sum_{k=1}^{n} h_k \frac{dy_k}{dt},$$

where h_k are partial molar enthalpies of species participating in reactions; y_k are their molar concentrations changing corresponding to kinetical equation:

$$\frac{dy_k}{dt} = \sigma_k , \; \sigma_k = \sum_{i=1}^{r} \nu_{ki} \varphi_i , k = 1,2,\ldots, n ,$$

where ν_{ki} is the stochiometric coefficient of k-species in the i-reaction; φ_i is the rate of i-th chemical reaction; r is the number of chemical reactions. When the reactions proceed with the participation of solid phases, kinetical functions should take into account the reaction retardation by a poorly permeable layer.

In general case, chemical reactions in the base are also possible.

At the initial time moment, we have $T = T_0, y_k = y_{k0}$, because initial coating contains some part of the reaction product formed during deposition.

This model does not take into account explicitly the structure transformations in the coating (porosity evolution and shrinkage due to porosity change), intermixture of melts of coating and substrate in the contact area, chemical reactions between them, mechanical stresses due to high temperature gradients and properties change in the reactions etc. However, this model can give a fair idea of the heat necessary for heating, synthesis regimes and basic phases appearing during the treatment, even if we use additional simplifications [8, 9]. Assuming that the coating is thermally thin in comparison with the boundary layer width that forms due to thermal conductivity during treatment time, and the thickness of the substrate of high thermal conductivity is lesser then its width and length, we take integral of the thermal conductivity equation with y-coordinate taking into account the boundary conditions. Additionally, if we believe that scanning thickness of electron beam covers the transversal size of the part (which provides the uniform heating in this direction, Fig. 1), we come to one-dimensional thermal conductivity equation:

$$c_b\rho_b\left[1+\frac{h_p c_p \rho_p}{h_b c_b \rho_b}\right]\frac{\partial T}{\partial t} = \frac{\partial}{\partial x}\left(\bar{\lambda}\frac{\partial T}{\partial x}\right) - \sigma\varepsilon_{eff}\left(T^4 - T_W^4\right) + \frac{1}{h_b}\Phi'(x,t) + \frac{h_p}{h_b}\cdot W(y_k, T), \quad (1)$$

where

$$\lambda_{eff} = \lambda_b\left(1+\frac{\lambda_p h_p}{\lambda_b h_b}\right), \ \varepsilon_{eff} = \left(\varepsilon_p + \varepsilon_b\right)/h_b, \ \Phi'(x,t) = q_{eff}\, exp\left[-\frac{(x-Vt)^2}{R_0^{\ 2}}\right].$$

For each system (Ti-Ni, Ni-Al, Al-Ti and etc.) we can write a detailed reaction scheme using state diagram and thermodynamical data. In this case, we should use kinetic equation system to describe the evolution of phase composition, similarly to [19, 20].

Here, we restrict the model by summary reaction:

«summary reagent» - «summary product»;

however, the kinetic function should account the reaction retardation by the layer of reaction product.

As a result, we come to the reduced model with one equation instead of kinetic equation system:

$$\frac{dy}{dt} = \omega = k\varphi(y)exp\left(-\frac{E}{RT}\right), \quad (2)$$

where y is the conversion degree; E is the activation energy for summary reaction; k is the pre-exponential factor, R is the universal gas constant; $\varphi(y)$ is the kinetic function with the form depending on micro-scale processes. Then, $W = Q\omega$, where Q is the summary chemical heat release. Similar approach is tried-and-true method during irreversible processes modeling in the combustion theory. It allows obtaining qualitative regularities on the basis of reduced model.

In this work, we assume $\varphi(y) = (1-y)exp(-my)$.

To carry out the qualitative analysis of the problem including equations (1) and (2) and boundary conditions $x = 0, \infty$: $\partial T/\partial x = 0$, we come to dimensionless variables

$$\theta = \frac{T - T_*}{T_* - T_0}, \ \tau = \frac{t}{t_*}, \ \xi = \frac{x}{x_*},$$

where $x_* = \sqrt{\lambda_b t_* / (c_b \rho_b)}$ is the spatial scale; the temperature scale T_* is determined as the temperature of product of solid-phase combustion $T_* = T_0 + \frac{Q}{c_p \rho_p}$; $t_* = \frac{c_p \rho_p R T_*^2}{EQk} exp\left(\frac{E}{RT_*}\right)$ is time for reaction completion under adiabatic conditions. Then the reduced model turns into

$$(1 + K_c \varepsilon)\frac{\partial \theta}{\partial \tau} = (1 + K_\lambda \varepsilon)\frac{\partial^2 \theta}{\partial \xi^2} + \frac{\varepsilon K_c}{\theta_0}\varphi(y)\psi(\theta) + S\, exp\left(-\frac{(\xi - \omega\tau)^2}{\delta}\right) - B\left[(\theta + \sigma)^4 - (\theta_W + \sigma)^4\right]; \quad (3)$$

$$\frac{\partial y}{\partial \tau} = \gamma \varphi(y)\psi(\theta);$$

$$\xi = 0: \ \frac{\partial \theta}{\partial \xi} = 0; \ \xi \to \infty: \ \frac{\partial \theta}{\partial \xi} = 0;$$

$$\tau = 0: \ \theta = -1, \ y = y_0,$$

where

$$\psi(\theta) = exp\left[\frac{\theta\theta_0}{1 + \theta/\sigma}\right];$$

$$\delta = \frac{R_0^2 c_b \rho_b}{\lambda_b t_*}, \ \gamma = \frac{c_p \rho_p R T_*^2}{EQ} << 1, \ \sigma = \frac{T_*}{T_* - T_0} > 1, \ \theta_W = \frac{T_W - T_*}{T_* - T_0} < 0; \ \theta_0 = \frac{T_* - T_0}{RT_*^2} E >> 1;$$

$$S = \frac{q_{eff} t_*}{h_b c_b \rho_b (T_* - T_0)}, \ B = \frac{(T_* - T_0)^3 t_* \sigma \varepsilon_{eff}}{c_b \rho_b}; \ K_\lambda = \frac{\lambda_p}{\lambda_b}; \ K_C = \frac{c_p \rho_p}{c_b \rho_b}; \ \varepsilon = \frac{h_p}{h_b} > 1; \ \omega = \frac{V t_*}{x_*}.$$

Model contains a large number of parameters. Depending on relations between them, various conversion regimes are possible.

The problem was solved numerically. The implicit difference scheme and double-sweep method were used.

4. Results of numerical investigation

We assume $\theta_W = -1$; $\sigma = 1.3$; $\delta = 1$.

If the coating is absent, $\varepsilon = 0$. In this case, we shall obtain:

$$\frac{\partial \theta}{\partial \tau} = \frac{\partial^2 \theta}{\partial \xi^2} + S\, exp\left(-\frac{(\xi - \omega\tau)^2}{\delta}\right) - B\left[(\theta + \sigma)^4 - (\theta_W + \sigma)^4\right]. \quad (4)$$

The scales for temperature and time are not determined here. We find them from the conditions of $S = 1$ and $\omega = 1$.

Temperature distribution for this variant of the model and for $\delta = 1$ is shown in Figure 2. For $B = 0$ (Figure 2a) and for $B \neq 0$ (Figure 2b), after initial non-stationary curve section, the quasistationary regime establishes in the system, when invariance of maximal temperature is typical for the process, the value of maxima depends on δ. However, quasistationary regime realizes when linear size of the part is larger than $\sqrt{\delta}$.

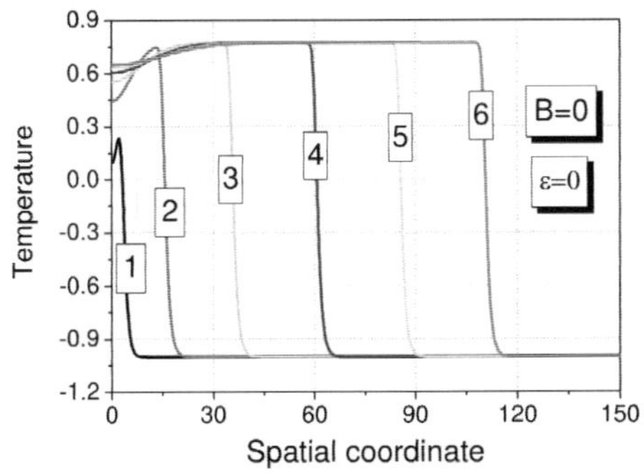

a

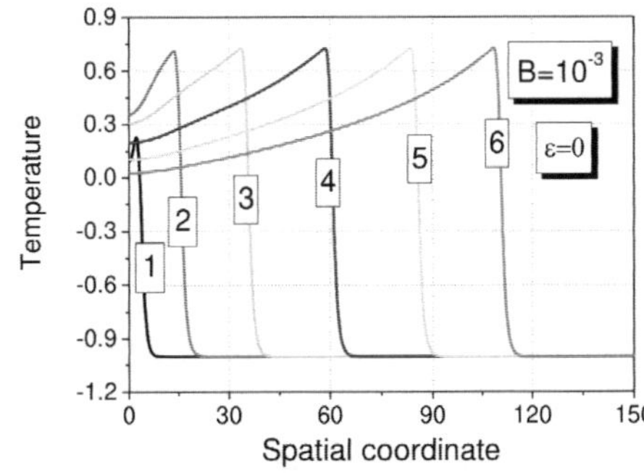

b

Figure 2. Temperature distribution (in a.u.) during substrate treatment for B = 0 (a) and B $\neq$ 0 (b); $\delta = 1$. Time (in a.u.): 1) 0.3; 2) 1.5; 3) 3.5; 4) 6.0; 5) 8.5; 6) 11.0.

The limiting variant of the model of reaction initiation in plane layer without substrate accounting can be obtained when

$$x_* = \sqrt{\lambda_p t_* / (c_p \rho_p)};\; S = \frac{q_{eff} t_*}{h_p c_p \rho_p (T_* - T_0)};\; B = \frac{(T_* - T_0)^3 t_* \sigma \varepsilon_{eff}}{c_p \rho_p};\; \varepsilon_{eff} = 2\varepsilon_p / h_p .$$

Then the thermal conductivity equation takes the form of

$$\frac{\partial \theta}{\partial \tau} = \frac{\partial^2 \theta}{\partial \xi^2} + \frac{1}{\theta_0} \varphi(y)\psi(\theta) + S\, exp\left(- \frac{(\xi - \omega\tau)^2}{\delta} \right) - B\left[(\theta + \sigma)^4 - (\theta_W + \sigma)^4 \right] \qquad (5)$$

and boundary conditions stay the same in any case.

Figure 3 demonstrates that, similarly to [8, 9], the chemical reaction starts at some distance from surface $\xi = 0$, and then the process continues in quasistaionary regime. When capacity density of the source decreases, the velocity of its travel along the surface increases; the reaction near the surface $\xi = 0$ is completed when external source leaves this area. The lower the capacity density of the effective external source, the lower the temperature and extension of the reaction zone.

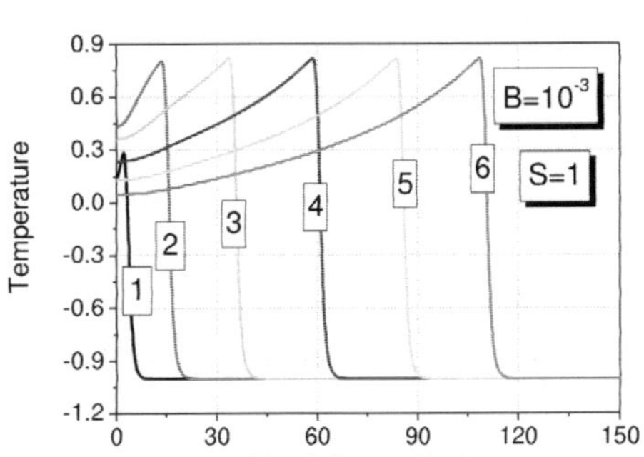

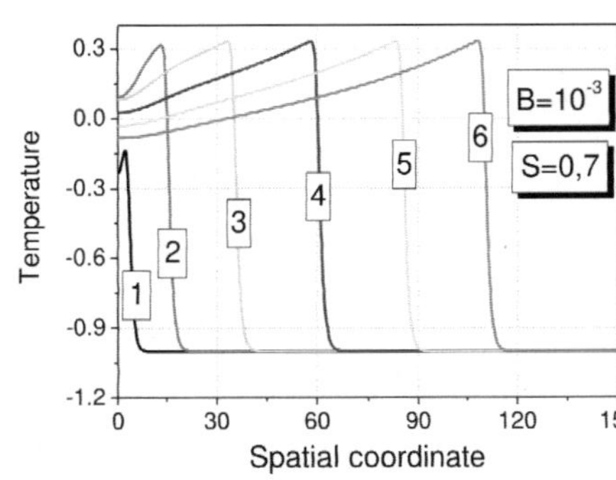

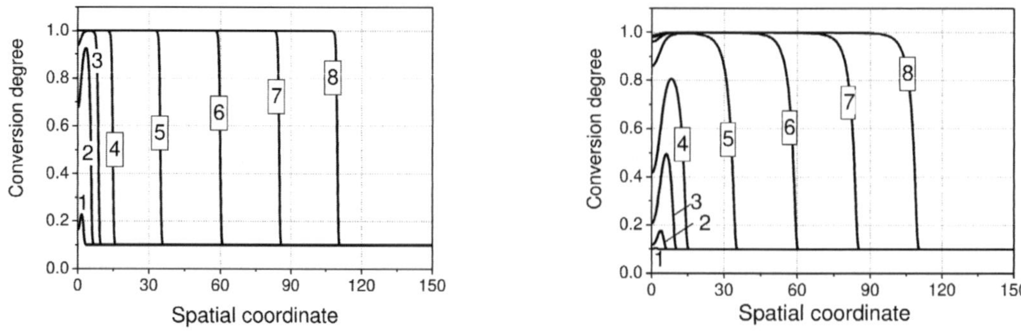

a b

Figure 3. Temperature and conversion degree distribution for different time moments during reaction initiation on thin layer without heat losses to substrate. Time for temperature curves coincides with the data from Figure 2. Time for conversion degree: 1) 0.3; 2) 0.6; 3) 0.9; 4) 1.5; 5) 3.5; 6) 6.0; 7) 8.5; 8) 11.0.

When synthesis is realized on the substrate, its properties affect the process characteristics.

On the one hand, the substrate requires the major part of the heat for heating. On the other hand, if the substrate is characterized by high thermal conductivity, it will promote the reagent heating in front of the moving external source. These effects were not taken into account in the simplest models described in [7].

The calculations have shown that the critical conditions exist in the problem. They separate the process of treatment accompanied by initiation of chemical conversions and the process of treatment without chemical reactions. Critical conditions were demonstrated by varying different parameters. Figure 4 shows different conversion degree when S changes. Different technological conditions give different composition of the coating. If we take into account the detailed reaction scheme, different coating composition will be expected in the numerical model varying the composition of green powders, similarly to [19, 20].

5. Conclusion

Since the Ti-Ni, Ni-Al and Al-Ti systems, especially in form of the coatings, are promising for practical implementation, the preliminary experimental data and suggested reduced model give a good start for further investigations. The model, despite the simplicity, reflects qualitatively basic regularities and will be used for prognostic model construction. The critical conditions should be taken into account to predict the technology conditions providing a given composition of the coating.

References

[1] Patil K C, Hegde M S, Rattan T and Aruna S T 2008 *Chemistry of Nanocrystalline Oxide Materials: Combustion Synthesis, Properties and Applications* (World Scientific)

[2] Maximilian Lackner 2010 *Combustion Synthesis: Novel Routes to Novel Materials* (Vienna: Bentham Science Publisher Ltd)

[3] Sychev A E and Merzhanov A G 2004 Self-propagating high-temperature synthesis of nanomaterials. *Russian chemical reviews* **73** 2 147-159

[4] Shipko A A, Pobol I L and Urban I G 1995 *Strengthening of steels and alloys using electron-beam heating (in Russian),* (Minsk: Nauka i technika) p 280

[5] Rykalin N N, Zuev A A and Uglov A A 1978 *The foundation of electron-beam treatment of materials (in Russian)* (Moscow: Mashinostroenie) p 239

[6] Schiller S, Heisig U and Panzer S 1995 *Electron Beam Technology* (Berlin: Technik GmbH)

IOP Conf. Series: Materials Science and Engineering **140** (2016) 012018 doi:10.1088/1757-899X/140/1/012018

[7] Shishkowskii I V 2009 *Laser synthesis of functional-gradient of meso structures and volume production (in Russian)* (Moscow: Nauka) p 422
[8]. Knyazeva A G, Pobol I L and Gordienko A I 2004 Coating formation in SHS-regime during thermal treatment of material by moving energy source. *7-th Int. Conf. on modification of Materials with Particle Beams and Plasma Flows, Tomsk, Russia*, 25-30 July 2004 pp 178-183
[9] Knyazeva A G, Pobol I L Gordienko A I., Demidov V N, Kryukova O N and Oleschuk I G 2007 *Physical mesomechanics* **10** 3-4 207-220
[10] Appel F, Heaton-Paul J D and Oehring M 2011 *Gamma Titanium Aluminide Alloys: Science and Technology* (New York: Wiley)
[11] Yong S, Xiaofei D, Fugang W, Tan Y and Yang J-M 2004 *Material Science* **39** 6583-6589
[12] Si-Young Sung, Young-Jig Kim 2007 *Intermetallics* **15** 468-474
[13] Liu K, Ma Y C, Gao M, Rao G B, Li Y Y, Wei K, Xinhua Wu and Loretto M H 2005 *Intermetallics* **13** 925-938
[14] Harding R A, Wickins M, Wang H, Djambazov G and Pericleous K A 2011 *Intermetallics* **19** 805-813
[15] Hu D, Wu X and Loretto M H 2005 *Intermetallics* **9** 914-919
[16] Shishkovsky I V, Kuznetsov M V, Morozov Yu V and Parkin I P 2004 *J. Mater. Chem.* 1**4** 344 – 344
[17] Mohammad Vaezi, Hermann Seitz, Shoufeng Yang 2013 *Int. J. Adv. Manuf. Technol.* **67** 1721 – 1754
[18]. Sikka V K 2000 *Intermetallics* **8** 1329–1337
[19] Knyazeva A G and Sorokova S N 2008 *Theoretical foundations of chemical engineering* **42** 443-451
[20] Sorokova S N and Knyazeva A G 2009 *Physical mesomechanics* **12** 77-90

Research on the possibility of restoring blades while repairing gas turbine engines parts by selective laser melting

V G Smelov[1], A V Sotov[1], A V Agapovichev[1,a]
[1]Samara State Aerospace University (National Research University), Moskovskoe sh. 34, Samara, 443086 Russia

[a]E-mail: agapovichev5@mail.ru

Abstract. We study the possibility of restoring the blades of chromium-nickel materials for the repair of parts of gas turbine engines using selective laser melting technology. The stages of preparation of the items to repair and reconditioning are considered in detail, the algorithm of the recovery process to a 3D machine has been developed. Chemical analysis of the raw material and facing material has been performed. Maps of distribution of chemical elements in the fusion zone of the starting material with the surfacing material have been acquired. In order to study the nature of alloying materials fractographic analysis of the places of fusion was performed. A map of distribution of chemical elements in the fusion zone was obtained.

1. Introduction

Restoration of parts to restore their functionality is important in many industries, including aviation [1]. About 70% of the total number of recovered parts of aircraft, made of high-alloy steels and aluminum alloys are parts with surface defects in the form of wear, cracking, corrosion and mechanical damage [2].

The blades of gas turbine engines are the most stressed parts, determining resource labor input and cost of the engine. The number of blades in modern engines reaches 2 ... 3,5 thousand, therefore their manufacture is carried out in a large scale or mass production.

The resource of the blades of military aircraft engines is - 500 ... 1000 hours, for civil aircraft - 10 ... 20 thousand hours. The cost and complexity of manufacturing of a set of blades is 20 ... 35% of the total cost and the complexity of the engine [3, 4].

During operation, blades are subjected to: stretching and bending due to centrifugal forces, bending and twisting under the influence of the gas flow to variable voltages from vibration loads, exposure to elevated and high temperatures (300 ... OS 600 for the compressor blades and 800..1200 OS for turbine blades), etc.

Analysis of damage to parts of aircraft, nonrenewable currently due to lack of methods to restore them, shows that about 70% of the total amount are parts with surface damage depth of 0.4 - 2.0 mm, most of which have not fulfilled their life and are defected due to wear [5].

To restore the geometric dimensions of the surface of the blade by selective laser melting (SLS) an algorithm was developed. It includes a number of stages (figure 1).

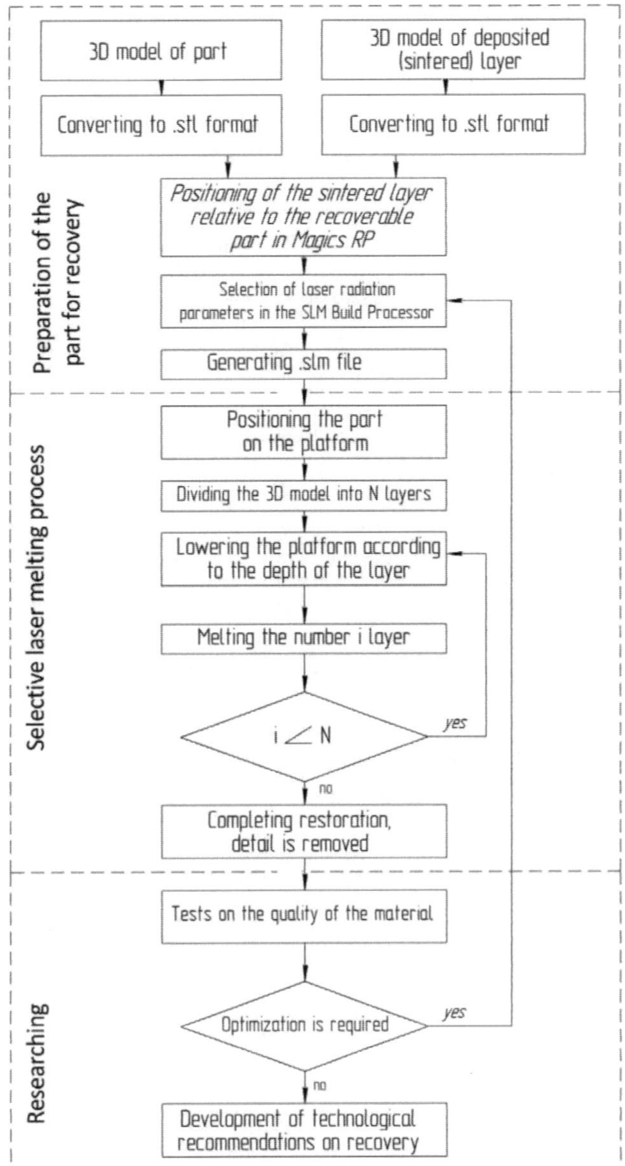

Figure 1. Algorithm for recovery of parts by SLM

Currently at different stages of design and production of new products widely used are methods of layering three-dimensional objects directly from their computer models, known as rapid prototyping technology.

Few areas of manufacturing three-dimensional objects layer by layer have been mostly developed, one of which is SLM technology of metal powder materials [6-8].

2. The methodology of the study

The object of research is a blade of compressor of gas turbine engine recovered by SLM method. Recovery of the end of the blade was carried on a 3D system of selective laser melting SLM 280HL (SLM Solutions GmbH, Germany). The system has a build camera with size of 280x280x350 mm and is equipped with an infrared fiber laser with a wavelength of 1075 nm and a maximum power of 400

W, operating in continuous mode. Growing of the samples was implemented in an inert gas (hydrogen).

Scanning electron microscope Tescan Vega was used for fractographic analysis. For the analysis of the elemental composition of the material X-ray spectral microanalysis was used.

Restoration of the end surface of the compressor blades using SLM technology consists of the following steps: creation of 3D models of the restored parts; creation of a 3D model of the sintered layer; Positioning the 3D model of the reconstructed detail on the construction platform; positioning the 3D model of the sintered layer with respect to the restored detail; preparation of the print file for 3D machine; fixing the recoverable detail ont the construction platform; positioning of the construction platform in the sintering chamber of 3D system SLM 280HL and the recovery itself.

At the first stage of recovery of the blades a 3D model of recoverable detail was created (figure 2a). Sintering layer is a material which is to be synthesized on the surface of the end of the blade [9]. Since the chosen vanes have the wear on the end face of the pen with understating blade height of 2 mm and further processing requires minimum machining allowance of 0.5 mm, the height of the sintered layer was 2.5 mm [10, 11]. Also, allowances of 0.5 mm were added to the 3D model from the trough and back of the blade (figure 2b). Blade profile measurement was performed on CMM [12]. Three-dimensional models have been developed in the software Siemens NX 8.5.

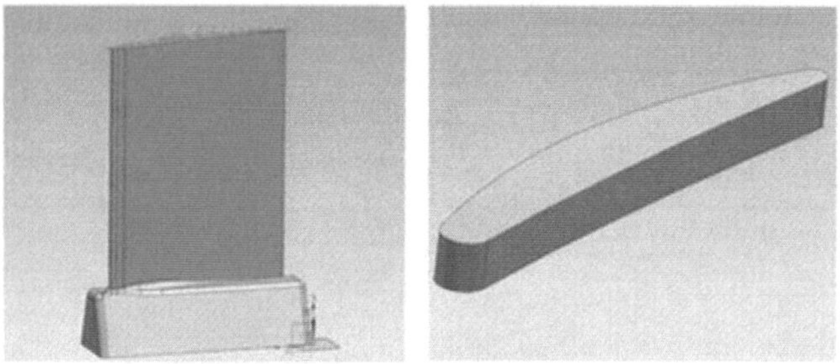

a) b)

Figure 2. 3D models of recoverable blade and fused lay; a) 3D model of restored blade; b) 3D model of the sintered layer

Positioning 3D model of the blade was performed using the software Magics RP, supplied by 3D machine SLM 280HL manufacturer. Positioning of 3D model of the blade was carried on construction platform of size of 100x100 mm (figure 3a).

Then the blade was mounted to the construction platform using laser welding. Positioning the blade on the platform must exactly repeat the location of the recoverable item on the virtual platform. Before that, the center from which the blade is positioned with grip at four points had been found on the platform (figure 3b).

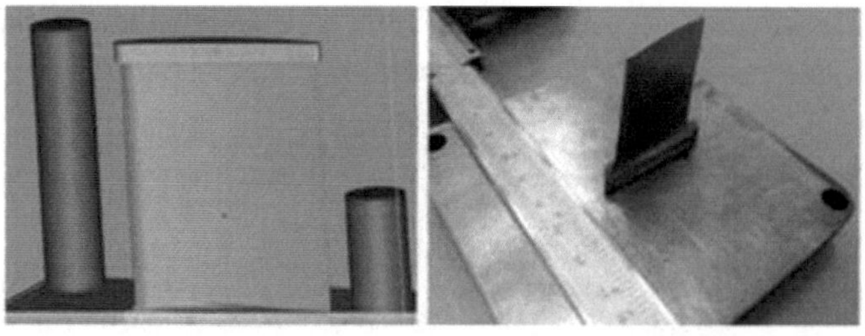

a) b)

Figure 3. Positioning of the blade on the construction platform; a) orientation of the model of the blade on the virtual platform; b) orientation of the model of the blade on the virtual platform

To position the construction platform in the fusion camera of 3D system SLM 280 HL it is necessary to establish the beginning of sintering of metal powder with laser, ie the first layer, which is to be grown on the blade end, will be the basis for the subsequent layers. To do this, the construction platform should be lowered down the piston at the height of the blade. The extremity end of the blade must coincide with the upper level of the well (figure 4a). Voids formed between the blade and the well with the use of the recoater are to be filled with powder until the layer is smooth. It took 1.87 kg of metal powder 316L to fill the volume (figure 4b).

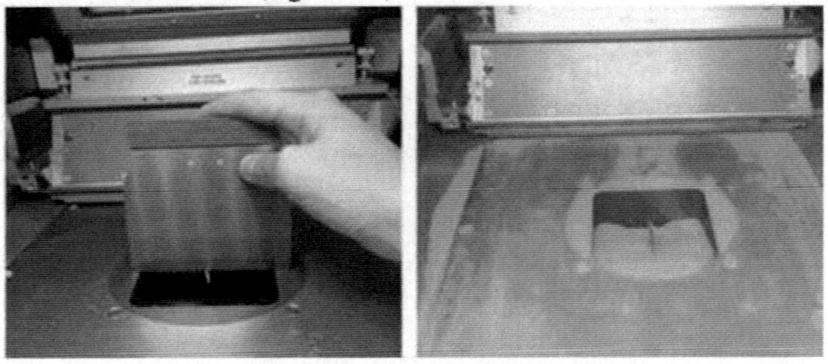

a) b)

Figure 4. Positioning of the blade on the construction platform; a) the search for zero of construction platform; b) filling the voids with metal powder

After all the preparatory work was carried out the recovery process of the compressor the blade by SLM was implemented. Restored blade is shown in figure 5.

Figure 5. Restored compressor blade.

3. Results and discussion

In order to study the microstructure of reduced layer fractographic studies using scanning electron microscopy method were performed in microsections obtained by electrical discharge machining [13]. Figure 6 shows the macro- and microstructure of the the blade in the recovery zone.

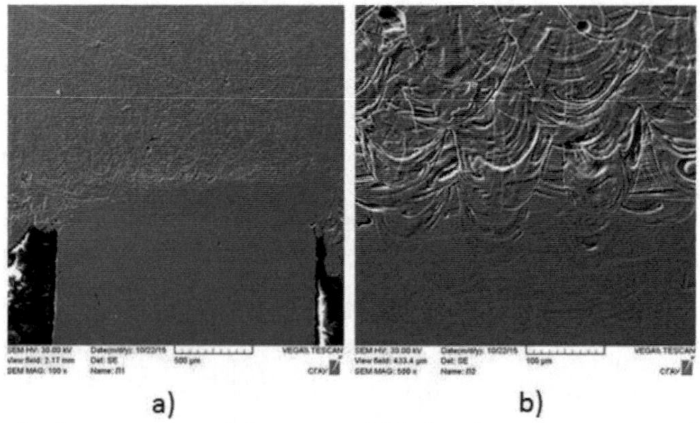

Figure 6. Macro and microstructure of the restored blade; a) macrostructure of the restored blade; b) microstructure of the restored blade

Figure 6 indicates that the border "basic material - surfacing" is clean adn there is a metallic bond (figure 6a). At a higher magnification of the reduction zone it is seen that at the border "core material - cladding" and at the surfacing material discontinuities are present, each up to 50 μm in diameter (figure 6b). Defects such as cracks were not found.

To study the border between the basic material and welding material a map of the distribution of chemical elements in the microsection was obtained (figure 7). Figure 7 shows that in the field of fusion "basic material - surfacing" a transition layer of width ~ 150 ... 200 μm was formed. The transition zone is not clearly defined by boundaries, indicating a good fusion of the base material with the surfacing material.

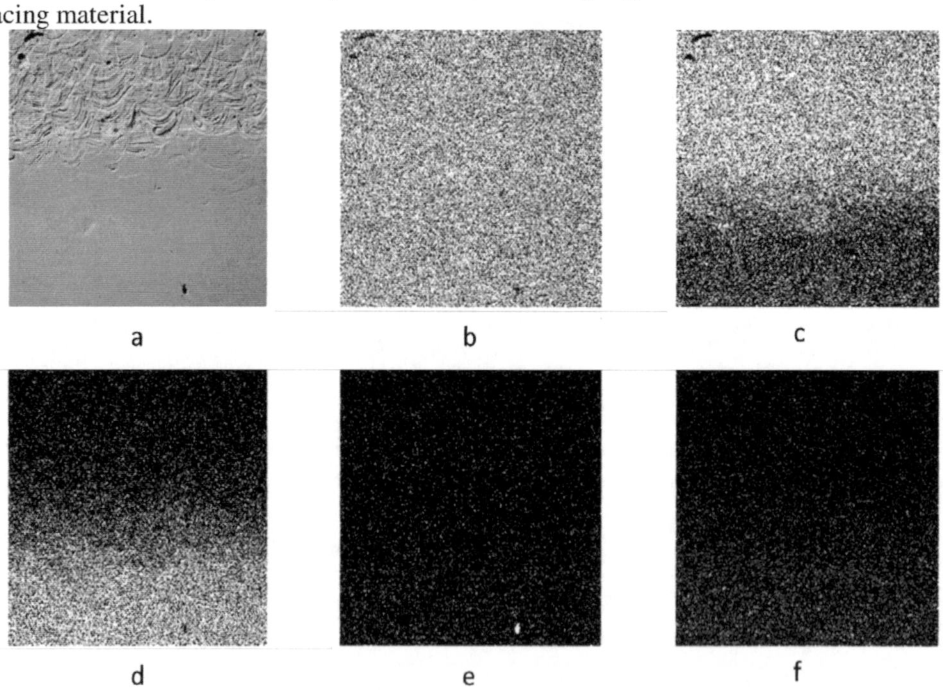

Figure 7. Macro and microstructure of the restored blade;
a) electronic image microsection (x500); b) Cr distribution; c) Fe distribution; d) Ni distribution; e) Al distribution ; f) Ti distribution

4. Conclusion

A study of the possibility of restoring a pen of a compressor blade worn during operations has been implemented. On the surface of the sintered layer lack of fusion has been detected, cracks have not been detected.

In all sections of the deposited layer of the blade boundary between the weld and the base material is not clear, there is a transition zone up to 200 μm thick.

Surfacing with SLM method allows to lower the amount of machining of the blades after recovery, since allowances for subsequent machining do not exceed 500 μm. It is possible to reduce the allowance for machining to 100 μm.

The developed method of process optimization by welding with SLM method allows to increase the efficiency of the recovery of parts of aircraft.

Further studies will focus on the impact produced by the main process parameters, such as laser power, scanning speed, thickness of the sintered layer, etc., on the porosity of the sintered layer, with a view to minimizing it.

To reduce the time spent on the preparation to the recovery process of parts by selective laser melting it is planned to design and manufacture devices for quick installation and positioning of parts.

Acknowledgements

These studies were conducted on the equipment of CAM technology common use center (RFMEFI59314X0003).This work was supported by the Ministry of education and science of the Russian Federation in the framework of the implementation of the Program "Research and development on priority directions of scientific-technological complex of Russia for 2014-2020".

References

[1] Barvinok V A, Semenov V G, Sotov A V and Kosirev S A 2014 Recovery of the end of the pen of a blade of turbine engine using pulsed laser deposition *J. Problems of mechanical engineering and automation* **3** 161-5

[2] Smelov V G, Sotov A V and Agapovichev A V 2016 Recovery Technology Features of Aerospace Parts by Layering Synthesis *J. Key Engineering Materials* **684** 316-22

[3] Makarov E L 2014 Theory of weldability of steels and alloys ed E L Makarov (Moscow: Publishing house of the MSTU named after Bauman) p 487

[4] Vdovin R A and Smelov V G 2014 Elaboration of a casting defects prediction technique via use of computer-aided design systems *International J. of Engineering and Technology (IJET)* **6(5)** 2269-75

[5] Nazarov A P 2013 Features of machine design for selective laser sintering *J. Vestnik MSTU "STANKIN"* **1** 76-79

[6] Landolfo R, Mammana O, Portioli F, Di Lorenzo G and Guerrieri M R 2008 Laser welded built-up cold-formed steel beams: Experimental investigations Thin-Walled Structures **46** 781–91

[7] Sufiiarov V Sh, Popovich A A, Borisov E V, Polozov I A 2015 Selective laser melting of heat-resistant nickel alloy *J. Tsvetnye Metally* **1** 79-84

[8] Grjaznov M U, Shoshin S V, Chuvildeev V N 2012 Effect of mesostructural hardening of steel 316L during stratified laser alloying *J. Bulletin of the Nizhny Novgorod University named after N. Lobachevsky* **5 (1)** 45-50

[9] Vdovin R A 2013 Multiproduct manufacture process improvement J. *Izvestiya of Samara Scientific Center of Russian Academy of Sciences* **6(3)** 612-19

[10] Khaimovich A I, Balaykin A V and Galkina N G 2016 Phenomenological modeling of rheological properties of polyetheretherketones reinforced with high modulus carbon in the machining process by a reversible analysis method *J. Key Engineering Materials* **685** 119–23

[11] Skuratov D L, Zhidyaev A N and Sazonov M B 2014 Solid carbide end mills tool life increase in titanium alloys machining by design development and rational choice of geometrical parameters *J. Research journal of applied sciences* **9(11)** 767-70
[12] Pechenin V A, Bolotov M A and Ruzanov N V 2014 Development of a method of ICP algorithm accuracy improvement during shaped profiles and surfaces control *International J. of Engineering and Technology* **6(5)** 2229-35
[13] Balaykin A V, Kondratev A I and Galkina N G 2016 Determination of rheological properties of titanium alloys under conditions of high strain rates *J. Key Engineering Materials* **684** 222-26

Phase Composition and Microstructure of Ti-Nb Alloy Produced by Selective Laser Melting

Yu P Sharkeev[1,2], A Yu Eroshenko[1], Zh G Kovalevskaya[2,1], A A Saprykin[3,1], E A Ibragimov[3,1], I A Glukhov[1], M A Chimich[1], P V Uvarkin[1] and E V Babakova[3]

[1] Institute of Strength Physics and Materials Science of the Siberian Branch of the Russian Academy of Sciences, 2/4, pr. Akademicheskii, Tomsk, 634021, Russia
[2] Tomsk Polytechnic University, Lenina avenue, 30, Tomsk, 634050, Russia
[3] Yurga Technical Institute of Tomsk Polytechnic University, 26, Leningradskaya Avenue, Yurga, 652055, Russia

E-mail: sharkeev@ispms.tsc.ru

Abstract. The phase composition and microstructure of Ti-Nb alloy produced from composite titanium and niobium powder by selective laser melting (SLM) was studied. Produced monolayered Ti-Nb alloy enhanced the formation of fine-grained and medium-grained zones with homogeneous element composition of 36-38% Nb mass interval. Alloy phase composition responded to β-alloy substrate phase (grain size was 5-7 μm) and non-equilibrium martensite α''- phase (grain size was 0.1-0.7 μm). α''-phase grains were found along β-phase grain boundaries and inside grains, including decreased niobium content. Alloy microhardness varied within 4200-5500 MPa.

1. Introduction

Bioinert low-modulus Ti-Nb alloys [1] are more preferable as medical alloys. Due to their physic-mechanical characteristics—mainly low elastic modulus and bioinert properties—these alloys are excellent materials for medical implants [1, 2]. It should be mentioned, that Nb-doped alloy including 40 wt% of Nb (Ti40Nb) exhibits the lowest elastic modulus for such a system (55-70 GPa*)* [1]. The elastic modulus for most alloys is within the range of 100-120 GPa [4, 6], which, in its turn, is significantly higher than in the case of bones (15-50 GPa). Based on the above-mentioned facts, an upcoming trend in medical materials science is the production of biocompatible titanium alloys with low elastic modulus, for example, alloys of Ti-Nb or Ti-Nb-Zr systems [6]. Traditional methods used in producing the titanium-niobium alloys are extraction metallurgical methods. Nevertheless, the application of traditional methods for producing titanium-niobium alloys with homogeneous microstructures is not always effective. There are also so-called methods of powder metallurgy, which could include additive manufacturing technology. Applying additive technology, for example 3D-printing mode selective laser melting (SLM) could be, in most cases, the only alternative method in manufacturing composite mold castings and/or parts using NC (numerical control) machine [3-5]. In this respect, the SLM method could be rather challenging in the production of implants and endoprostheses. Another important aspect is that this method produces items with specified porosity, which, in its turn, improves the implant material to bones osseointegration. This is today's current

IOP Conf. Series: Materials Science and Engineering **140** (2016) 012020 doi:10.1088/1757-899X/140/1/012020

issue involving both the development of a SLM method and the study of phase composition and structure formation. This paper describes the investigation of specific features of alloy structure formation and phase composition of Ti40Nb produced by SLM.

2. Experimental Procedure

Ti-Nb alloy samples produced by SLM using VARISKAF 100mV were investigated [6-8]. As the melting temperature of titanium (1725°C) and niobium (2468°C) are significantly different, composite titanium and niobium powder (60 wt% Ti and 40 wt% Nb) with average particle size of 20 μm was used as the reference material. Powdered material was sifted on VT1-0 substrate as a thin layer with homogeneous thickness and density. The following SLM modes were selected: laser radiation power, 105W; laser beam scan speed, 2000-3000 mm/min; laser beam diameter, 0.4 mm; scan step, 0.1 mm; substrate preheating, 200 °C. The scanning direction of each succeeding layer was altered by 90°. To produce a monolayer, the composite powdered layer thickness was 0.5 μm, determining the melt layer thickness and deleting the monolayer from the substrate.

Optical microscopy X-ray phase analysis, SEM, TEM and microhardness testing were performed using DRON-7, LEO EVO 50, AXIOVERT-200MAT, JEM-2100 and Duramin-5 (SKC "NANOTECH", ISMPS SB RAS, Russia).

3. Results and Discussion

Molding of bulk sample by SLM method was based on scanning along a predetermined trajectory by melting thin powder layer and crystallizing the material on interlaying substrate tracks. Multi-layered sample surface topography is identical to a monolayer, as the molding of each succeeding layer in the sample replicates the previous monolayer. In this case, the molding of a monolayer in the first cycle is the most important phase and determines the entire homogeneous structure-phase molding of the bulk produced sample. The total thickness of melted monolayer amounts up to 300 μm.

SEM investigation of sample cross-sections showed that the bulk sample is molded as a result of the superposition of the layers, which could be described as monolayer (Figure 1). Such a monolayer embraces completely melted and crystallized material. There are two zones within the monolayer. The first one involves fine-grained structure with grain size of 2 to 8 μm, further merging into medium-grained structure (zone 2) within an interval of 7 to 24 μm. Pores can be observed across the monolayer section.

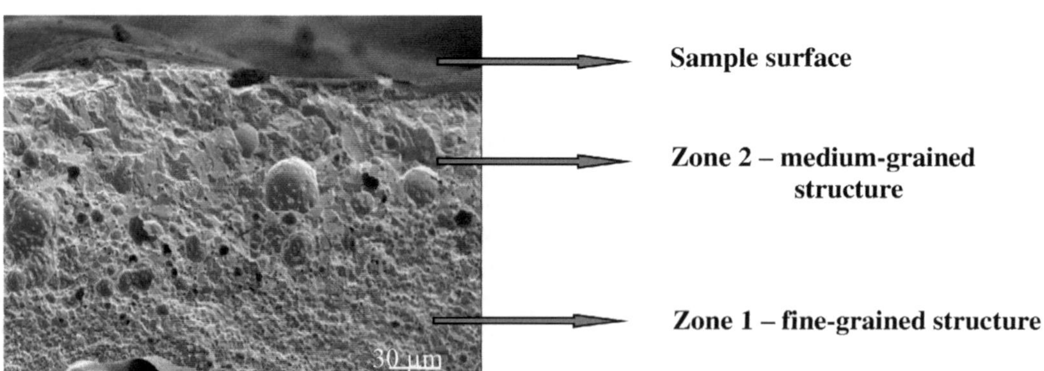

Figure 1. SEM image of fracture surface

In terms of zone structure aspects, the monolayer molding could be described as follows. Under exposure of a laser beam, the powdered material in the vicinity of the heating source, i.e. upper powdered layer, is melted. This results in the formation of lower spongy monolayer zone. During cooling, heat dissipation is evident in the zones of sintered powder and substrate in the upper melt material. Inversely from the boundaries, the sintered powdered material crystallizes forming clowholes. In the lower crystallizing bath, the increase of cooling rate furthers the formation of fine-grained structure with inclusions of small gas pockets along the grain boundaries. In the upper crystallizing bath, where heat dissipation is impeded in the substrate, cooling provides rather balanced material crystallization conditions. In this case, a medium-grained structure zone with average porosity and large grains is formed in the upper monolayer. Between the upper and lower crystallization zones large rounded pores emerge. As a result of layering, the translation of the monolayer structure itself occurs within the monolayer sample.

Thus, it can be stated that the formation of distinct fine-grained and medium-grained structured zones in the sample section is related to different temperature-time conditions of powder melting, crystallization and cooling in selective laser melting. It should be noted that monolayer microhardness in axial section ranges from 4200 to 5500 MPa.

Regardless of the non-uniform structure of the monolayer, titanium and niobium are distributed evenly throughout all the zones, as indicated by the mapping results of fractures by SEM-EDST (Figure 2). Quantitatively, the basic melt components are insignificant, 36-38 wt% of Nb.

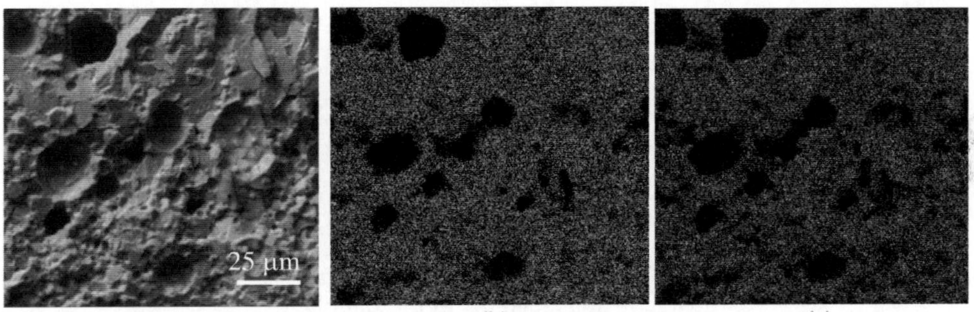

(a) (b) (c)

Figure 2. SEM images of fracture (a); Ti distribution (b); Nb distribution (c)

Optical image of zone 1 with fine-grained structure is depicted in Figures 3a and 3b. Basically, the microstructure embraces equiaxial grains of 5 µm. Grain boundaries are decorated with less than 1 µm grains of different phase (Figure 3a). These tabular grains are characteristic for the martensite phase (Figure 3b). These samples can be identified as reflections from the diffracting plane of β-BBC (body-centered cubic) of Ti and Nb solid solution and non-equilibrium α"- phase with orthorhombic lattice on X-ray diffractograms (Figure 3c). The existing α"-phase is associated with the unbalanced cooling and crystallization conditions of composite powder [9, 10].

TEM method was used to investigate the microstructure of produced melt samples. Light field images of Ti40Nb microstructure sample with corresponding microdiffractions are depicted in Figures 4a and 4b.

Rather small grains of 0.1-0.7 µm are located on the β-alloy grain boundaries, which in turn corresponds to ultra-fine grained structure (Figure 4b). These tabular grains indicate the fact that they were formed in unbalanced conditions during cooling of melted powder. Microdiffractions show reflections of metastable titanium α''-phase (orthorhombic crystalline lattice), which could be characteristic for possible martensite phase formation [10]. It should be noted that α''-phase grains include low content of Nb of 13-19 wt%. Reflexes of α''-phase for zone axis $[\bar{1}01]_{\alpha''}$ were identified on the microdiffraction.

IOP Conf. Series: Materials Science and Engineering **140** (2016) 012020 doi:10.1088/1757-899X/140/1/012020

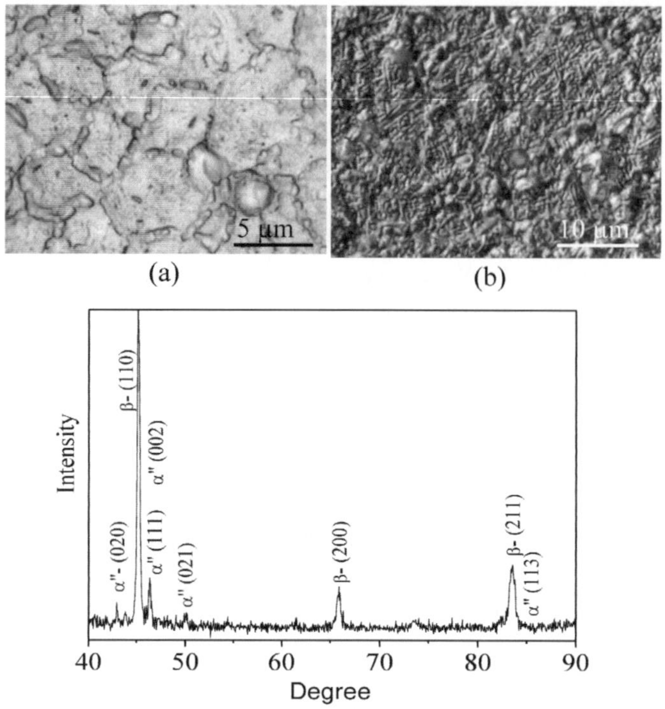

Figure 3. Optical images of the microstructure (a, b) and X-ray diffractogram fragment of the sample (c)

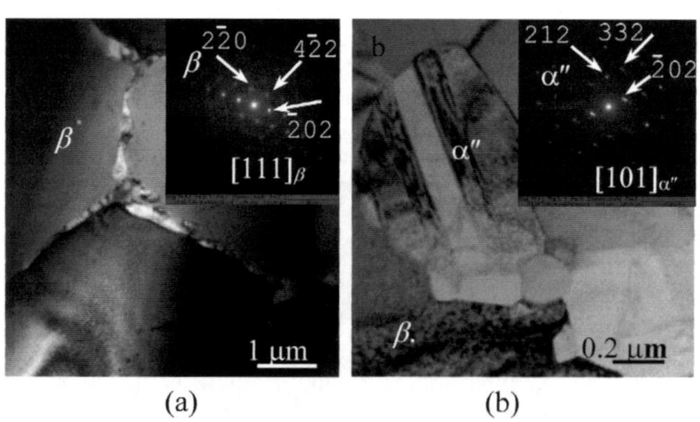

Figure 4. Light-field electron-microscopic images with corresponding microdiffractions of β-phase (a) and α''- phase (b)

Consequently, SLM method leads to molding of Ti-Nb alloy monolayer from composite powder of homogeneous element composition, which in its turn includes medium-fine grained structure zones. Forming highly homogeneous microstructure of bulk samples with specific geometrical dimensions requires specific conditions for powder cooling and crystallization. In this case, the monolayer and the item itself will have identical microstructures embracing rather fine grains and insignificant porosity.

4. Conclusions

Ti-Nb alloy monolayer with homogeneously distributed titanium and niobium was produced from composite titanium and niobium powder by SLM method. This monolayer embraced upper zone of fine-grained structure and lower zone of medium-grained structure. Average Nb content in the alloy within the monolayer along and across the sample changes insignificantly 36-38 wt%. Ti-Nb alloy includes two-phase state, i.e. base β-phase (grain size, 5-7 μm) and non-equilibrium martensite α''-phase (grain size is 0.1-0.7 μm), located on grain boundaries of β-phase. Grain of β-phase involves high Nb content (up to 40 wt%). Decreased Nb content (up to 20 wt%) is observed in grains of α''-phase. Formation of fine-grained and medium-grained structure zones are related to different temperature time conditions of crystallization and cooling of a layer melted during SLM.

Acknowledgment

Work was supported by the Russian Science Foundation, project No. 15-19-00191. The authors gratefully thank A I Tolmachev for his contribution in this research.

References

[1] Mitsuo Niinomi, Masaaki Nakai, Junko Hieda 2012 *Acta Biomaterialia* **8** 3888-3903
[2] Mohamed Abdel-Hady Ceepreel, Mitsuo Niinomi 2013 *J. of the Mechanical Behavior of Biomedical* **20** 407-15
[3] Zhao D, Chang K, Ebel T, Nie H, Willumeit R and Pyczak F 2015 *J. of Alloys and Compounds* **640** 393–400
[4] Zhuravleva K, Bönisch M, Prashanth KG et al. 2013 *J. Materials* **6**. 5700–12
[5] Shishkovsky I V 2009 *Laser synthesis of functional mesostructure and bulk products* (Moskow: FIZMATLIT)
[6] Saprykin A A, Ibragimov E A and Yakovlev V I 2014 *J. Applied Mechanics and Materials: Scientific* **682** 143–7
[7] Saprykin A, Saprykina N, and Krovina O 2008 *Materials Science Forum Vols. Trans Tech Publications, Switzerland* **575-578** 654-9
[8] Saprykina N A, Saprykin A A and Shigaev D A 2011 *Obrabotka metallov* **4(53)** 78-83
[9] Mantani Y and Tajima M 2006 *Materials Science and Engineering A.* **A 438–440** 315-19
[10] Colings E W 1984 *Physical Metallurgy of Titanium Alloys* (Metals Park, OH: American Society for Metals, Physical Description)

Mathematical modelling of thermal and kinetic phenomena in electron-beam technologies

A G Knyazeva[1,3], I L Pobol[2], V N Demidov[1]
[1] Tomsk Polytechnic University, Tomsk, Russia
[2] Physical-technical Institute of the National Academy of Sciences of Belarus, Minsk
3 Institute of Strength Physics and Materials Science of the Siberian Branch of the Russian Academy of Science, Tomsk, Russia

E-mail: anna-knyazeva@mail.ru

Abstract. The article presents a scheme for constructing models using kinetic laws for additional parameters. The work describes the example of the model of electro-beam treatment of a material with a coating. The study uses the simple kinetic law for powder layer evolution due to shrinkage. The model takes into account the melting of powder layer and substrate. The numerical solution gives the temperature field, evolution of the molten pool, the heat affected zone and the surface relief for different moments of time. The results depend on the treatment rate and electron beam energy.

1. Introduction

Electron-beam technologies are interesting for development of manufacturing and for investigation in the directions involving the modification of material properties and its surfaces [1–3]. Recent applications are connected with additive manufacturing [4]. The viability of laser-based and electron-beam-based additive manufacturing has been successfully demonstrated for titanium and nickel-based alloys, for stainless steel. etc. Thermophysical phenomena determining the additive manufacturing results are the same [5] that take a place in the processes of laser and electron beam welding, build up welding and surface treatment [6].

Additive methods for producing metallic parts are developed in the directions of powder and wire employment as green materials. There are technologies, where layer build-up happens due to the feeding of a wire or powder into the heated zone or into the melt formed by laser or electron beams. Another way of additive layer manufacturing using metal powder compositions uses the fusing of layers by scanning beam of small diameter. Note that electron-beam additive manufacturing is characterized by the building-up velocity of 120 cm^3/h, which is higher than 2-20 cm^3/h for similar laser machines. In addition, sintering of chemically active metals (particularly, titanium and their alloys) in vacuum is cheaper than that for corresponding laser process in protective gas environment.

In additive manufacturing applications, it is necessary to know and predict the possible shrinkage of a powder layer during Selective Laser or Electron Beam Melting. This phenomenon in different technologies can be linear [7] or nonlinear in connection with nonlinear thermal conductivity or phase transitions [8]. There are many physical factors that affect densification of products during sintering [8, 9]. These processes play a significant role when homogeneity of mechanical and physical properties of synthesized material is required in the direction normal to the directions of heated zone propagation and article production. In addition to the modelling of shrinkage during solid freeform fabrication [10] based on FEM (Finite-Element Method), it is important to study the models, where the shrinkage process can be described by kinetic laws based on physical regularities that are typical for them [11, 12].

Because the thickness of consecutive powder layers is small in comparison with specific heat scale and detail size and one powder layer includes only several particles, the approach schematically depicted in Figure 1 is methodologically correct and will yield prognostic models.

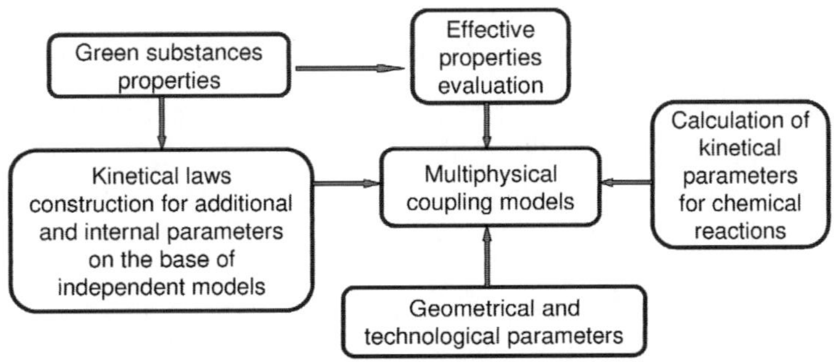

Figure 1. Conditional scheme for coupling models construction

Note that the construction of kinetic laws is possible at meso level, when the observation after individual particle behaviour is ensured. Here, the Phase-field method [13] and lattice Boltzmann equation method [14] are suitable. Unfortunately, the authors of similar works do not suggest the generalized formulae. Kinetic parameters for chemical reactions can be found using thermodynamics, quantum chemistry, molecular methods or experiment. Effective properties are quite structure-sensitive, and there are numerous theoretical approaches in micromechanics and thermal physics for their calculation.

Here, we demonstrate some features of the models with kinetic laws for evolution of powder layer thickness.

2. Thermophysical model formulation

A mathematical model of surface treatment of the plate (with preliminarily deposited powder layer) by scanning electron beam developed in the line can be presented in the form [15].

$$(c\rho)_{eff}\frac{\partial T}{\partial t}=\frac{\partial}{\partial x}\left(\lambda_{eff}\frac{\partial T}{\partial x}\right)+\frac{\partial}{\partial y}\left(\lambda_{eff}\frac{\partial T}{\partial y}\right)+\frac{q_{eff}}{h_b}f(x,y,t)-\sigma\varepsilon_{eff}\left(T^4-T_W^4\right) \quad (1)$$

$$x=0,H_x:\ \frac{\partial T}{\partial x}=0,\ y=0,H_y:\ \frac{\partial T}{\partial y}=0, \quad (2)$$

$$t=0:T=T_0, \quad (3)$$

where

$$f(x,y,t)=exp\left[-\frac{(x-Vt)^2}{R_0{}^2}\right],|y|\le h_y,\ f(x,y,t)=0,|y|>h_y,$$

h_y is the scanning width, H_x, H_y are the sizes of the plate along x and y directions, R_0 is the electron beam radius; σ is Stefan-Boltzmann constant; $\varepsilon_{eff}=(\varepsilon_1+\varepsilon)/h_b$ is effective emissivity; q_{eff} is the effective source power density of (with regard to volume absorption). Effective thermophysical properties are averaged along the thickness of the layers

$$(c\rho)_{eff} = c_b\rho_b + c_p\rho_p\, h_p / h_b \,,$$

$$(\lambda)_{eff} = \lambda_b + \lambda\, h_p / h_b \,,$$

The heat capacities of the substrate $c_b\rho_b$ and powder layer $c_p\rho_p$ change according to law

$$c\rho = \begin{bmatrix} c_L\rho_L, T \geq T_m \\ c_s\rho_s, T < T_m \end{bmatrix} + L_m\rho_s\delta(T - T_m) .$$

In general case, the thermal conductivity coefficients also depend on temperature.

Analyzing the calculation results for $T(x, y, t)$, one can investigate the evolution of melting pool and heat affected zone, which has been determined using conditionally given temperature (for example, $T = 700 - 900$ K). If the temperature does not reach this value, the visible change in the material structure is not observed.

Certainly, this model is correct when the coating (or powder layer) is absent.

However, phase composition determined on the base of phase diagrams using temperature value found from pure thermal conductivity problem will be quite approximate. Correctness of the approach [16], where the results of thermal problem solution are used to determine the phase structure by the phase-field method is ambiguous from the methodological point of view.

Additional complexification of thermal models conditioned by the consideration of the temperature dependencies of the properties also leads to temperature change, variation of heat affected zone and melting pool shape. The results do not change qualitatively in three-dimensional problems. New capabilities appear for determination of the melting depth and temperature distribution along the third axis for massive parts.

Though, thermal-physical models can be efficient due to taking into consideration various kinetical phenomena, which lead to evolution of properties.

3. Formal kinetics of shrinkage

The formation of phases, porosity and evolution of defects, change of properties, shrinkage are kinetical processes. To take into account these processes while models are constructed, it is necessary to formulate the kinetical laws. Processes at micro- and meso-levels should be affected by kinetical parameters and kinetical functions.

For example, to ensure the required properties for preliminarily deposited powder coating, the additional thermal treatment is necessary that leads to densification and shrinkage of this layer. From physical point of view, the shrinkage is connected with melting and mass transfer in different phases, with adhesive forces appearance between individual particles etc. All mentioned processes are thermally activated and proceed at micro- and meso-levels: in the volume of individual particles, in pores between particles, at the interfaces. We assume that the independent investigations were made for constructing kinetical law. As a result, we come to the kinetic equation

$$\frac{dh}{dt} = \varphi(\rho_p(T), h) = \varphi(\Theta(T), h),$$

where Θ is porosity.

The simplest form of the kinetic equation is

$$\frac{dh_p}{dt} = \begin{cases} -k_p\left(\dfrac{h_p}{h_{min}} - 1\right)^s exp\left(-\dfrac{E}{RT}\right), & h_p > h_{min}; \\ 0, & h_p > h_{min}, \end{cases} \quad (4)$$

$$t = 0: \quad h_p = h_{p0}, \tag{5}$$

where h_{min} is the minimal thickness of the coating obtained from powder layer with apparent density ρ_{p0}; k_2 is the constant of shrinkage rate; E is shrinkage activation energy (its value is determined by the processes limiting the shrinkage at micro-level), s is index in the kinetic function (it can be a whole or fractional number, depends for example, on particle surface shape). All parameters can be found from experiment or on the base of special physico-mathematical models. Lattice Boltzmann Equation method is suitable here [14, 17].

In the simplified approximation, we believe that thickness of powder layer is inversely proportional to density ρ_p. Because $\rho_p = \rho_{Me}(1-\Theta)$, we can write

$$\Theta = 1 - (1-\Theta_0)h_{p0}/h_p,$$

where Θ_0 is the porosity of poured powder layer of thickness h_{p0}.

To demonstrate the additional possibilities appearing in the model, taking into account the shrinkage, we assume that the substrate is made from iron ($c_{bs} = 0.825, c_{bL} = 0.608$ J/(g·K), $\rho_{bs} = 7.61$, $\rho_{bL} = 7.87$ g/cm³, $\lambda_b = 0.45$ J/(cm·sec·K), $T_{bm} = 1810$ K, $L_{bm} = 247$ J/g; $h_b = 0.5$ cm). The powder layer has the initial thickness of $h_p(t=0) = h_{p0} = 0.01$ cm and the following properties: $\lambda_p = 0.08$ J/(cm·sec·K), $c_{bs} = c_{bL} = 1.25$ J/(g·K), $\rho_{ps} = \rho_{pL} = 3.75$ g/cm³, $T_{pm} = 2318$ K, $L_{pm} = 389$ J/g.

Model will allow describing, in a first approximation, the surface relief change due to shrinkage during surface treatment [15].

We assume that $H_x = 4.0$; $H_y = 3.0$ cm, and properties are assumed constant (except heat capacities near the melting temperatures).

Kinetic parameters ($E = 100$ J/mol, $h_{min} = 0.005$ cm, $k_p = 5 \cdot 10^2$ cm/sec) were chosen to illustrate the shrinkage. Shrinkage observed experimentally can be 30-40%.

For $q_0 = 1 \cdot 10^5$ J/(cm²·sec) and for $V = 1{,}0$ cm/sec we get the following picture.

First, temperature in the heated zone does not reach melting temperature for powder layer (Figure 2a, left picture), but a small shrinkage takes place (Figure 2a, right picture). Even for quasi-stationary process (when maximal temperature does not change), the thickness of powder layer changes nonuniformly. If the temperature cycles at the points near the surface are different (Figure 3a), so in other points the temperature cycles are the same for different points along the X-axis. However, the shape of heat affected zone and of the area with temperature higher then T_{bm} change in time during the treatment. The pictures of the first line in Figure 4 illustrate this. The second line in the figures gives information about surface relief evolution during treatment.

If $q_0 = 1.5 \cdot 10^5$ J/(cm²·sec) and $V = 1.0$ cm/sec, we obtain other results. Maximal temperature during treatment is higher than in previous case (Figure 2b), more substantial shrinkage is observed. This is shown in Figures 2 and 4.

Temperature changes similarly with time. However, we can see here the temperature slope connected with substrate melting more clearly (Figure 3b). When temperature exceeds the melting temperature of powder material, we can see the molten pool in addition to the heat affected zone. Isolines in Figure 4 (a and c) correspond to temperature 350, 700, 1800 and 2300 K.

Calculations show that the relief of the coating after treatment depends on all technological parameter and on the initial thickness of the powder layer.

For practical applications, this model can be used to elaborate the recommendations for suitable parameter selection. By varying the treatment velocity, electron beam power, scanning thickness, etc., one can obtain enough uniform surfaces that ensure uniform properties.

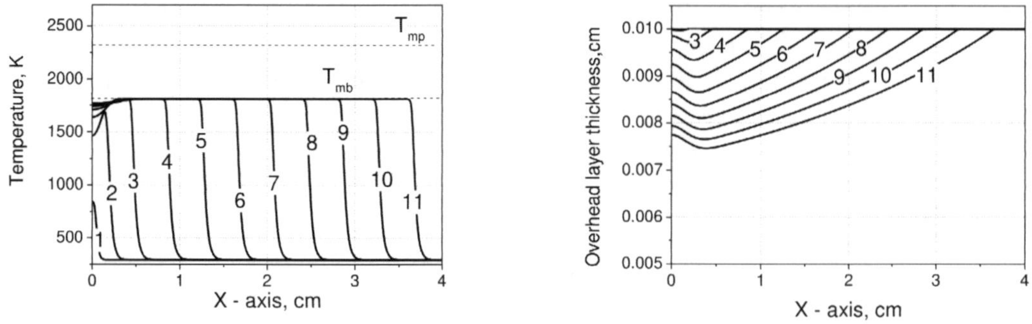

a)

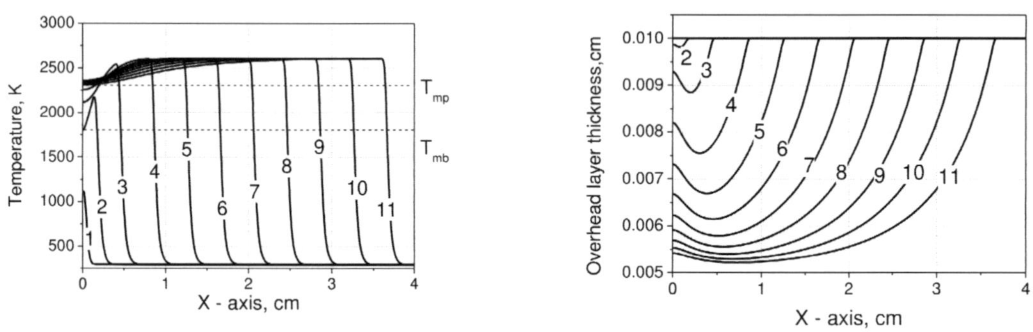

b)

Figure 2. Temperature and coating thickness along X-axis.
a) $q_0 = 1 \cdot 10^5$ J/(cm^2·sec); b) $q_0 = 1.5 \cdot 10^5$ J/(cm^2·sec)

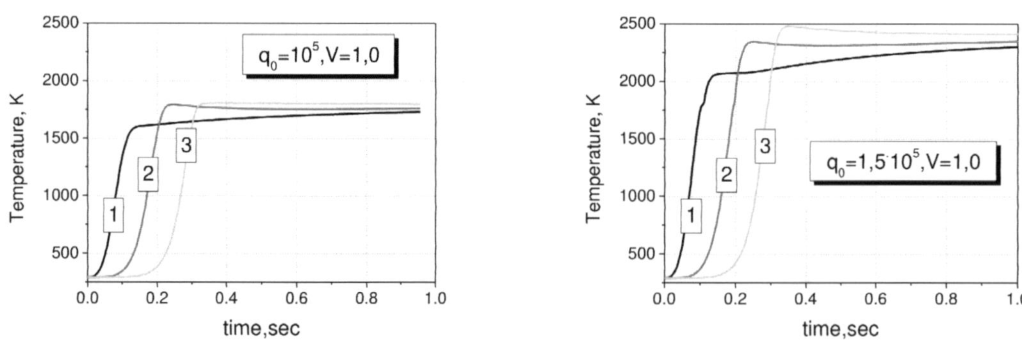

a) b)

Figure 3. Temperature in time for three points along X-axis: 1) $x = 0.1$; 2) $x = 0.2$; 3) $x = 0.3$ cm

4. Conclusions

The results demonstrate that thermal-kinetic model qualitatively describes the observed regularities: temperature growth with effective flux density increase, source velocity reduction, thickness change during shrinkage. This allows predicting the thickness change during treatment after kinetical parameter identification. Richer opportunities appear when chemical reactions have been taken into account. Some examples are presented in [11, 15].

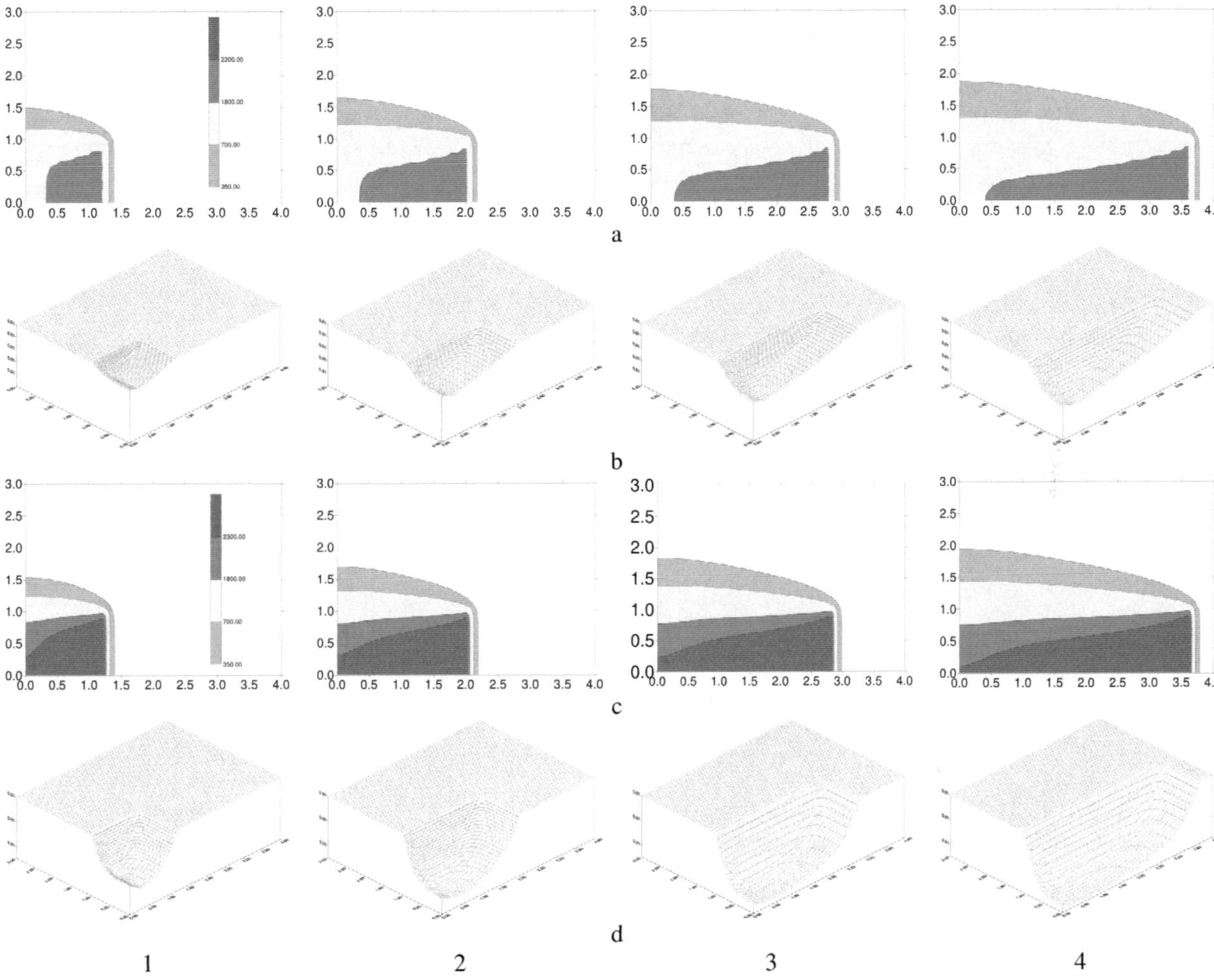

Figure 4. Evolution of temperature field (a, c) and thickness of powder layer (b, d) in time for $q_0 = 1 \cdot 10^5$ J/(cm^2·sec) (a and b) and $q_0 = 1.5 \cdot 10^5$ J/(cm^2·sec) (c and d). $t =$ 1) 1.25 sec; 2) 2.05 sec; 3) 2.85 sec; 4) 3.65 sec.

Acknowledgments
This work was supported by the Ministry of Science and Education of the RF, project No. 11.815.2014/K.

References

[1] Herderick E 2011 Additive Manufacturing of Metals: A Review *J. Materials Science and Technology Conf. and Exhibition (MS&T 11) (16-20 October 2011 Columbus, Ohio, USA)* vol 2 pp 1413-25

[2] Poate J M, Foti G and Jacobson D C 1983 *Surface Modification and Alloying: by Laser, Ion, and Electron Beams* (New York: Plenum Press) p 414 (DOI 10.1007/978-1-4613-3733-1)

[3] Shipko A A, Pobol I L and Urban I G 1995 *Strengthening of steels and alloys using electron-beam heating* (in Russian) (Minsk: Nauka i technika) p 280
[4] Gong X, Anderson T and Chou K 2014 *Manufacturing Rev.* 2 **1** 1–12
[5] Thomas M, Baxter G J, Todd I 2016 *Acta Materialia* **108** 26 - 35
[6] Rykalin N N, Zuev A A and Uglov A A 1978 *The foundation of electron-beam treatment of materials* (in Russian) (Moscow: Mashinostroenie) 239 p.
[7] Sercombe T B and Hopkinson N 2006 *Advanced engineering materials* 4 **8** 260–4
[8] Ranogajec J, Djuric M, Radeka M and Jovanic P 2000 *Ceramics − Silikáty* 2 **44** 71–7
[9] Ade M and Haußelt J 2003 *Journal of the European Ceramic Society* **23** 1979–86
[10] Dai K and Shaw L 2005 *Acta Materialia* 18 **53** 4743–54
[11] Knyazeva A G, Pobol I L Gordienko A I., Demidov V N, Kryukova O N and Oleschuk I G 2007 *Physical mesomechanics* 3-4 **10** 207-20
[12] Knyazeva A G and Bookrina N V 2003 *Proc. of 12th Int. Conf. on radiation physics and chemistry of inorganic materials (23-27 September Tomsk, Russia)* pp 167-70
[13] Sahoo S and Chou K 2016 *Additive manufacturing* **9** 14-24
[14] Attar E and Korner C 2009 *J. of Colloid and Interface Science* **335** 84-93
[15] Knyazeva A G 2009 *Mathematical modeling of the systems and processes. Special issue of Vestnik PerGTU* (in Russian) **17** 66 – 84
[16] Krivilyov M D, Kharanzhevskii E V, Lebedev V G, Danilov D A, Danilova E V and Galenko P K 2013 *The Physics of Metals and Metallography* 10 **114** 799-820
[17] Gürtler F-J, Karg M, Dobler M, Kohl S, Tzivilsky I and Schmidt M 2014 *Proc.* of *Int. Solid Freeform Fabrication (SFF) Symp. - An Additive Manufacturing Conf. (4-6 August Austin, Texas, USA)* pp 1099-117

IOP Conf. Series: Materials Science and Engineering **140** (2016) 012022 doi:10.1088/1757-899X/140/1/012022

Numerical analysis of initial stage of thermal shock

V N Demidov
Tomsk Polytechnic University

E-mail: vn_demidov@mail.ru

Abstract. The paper studies a problem of a thermal shock at the surface of a half-space, which properties are described by elastic-plastic model taking into account dynamic effects, heat inertia, coupling between thermal and mechanical fields. The problem is solved numerically using finite-difference method of S.K. Godunov.

1. Introduction

Numerous advanced technological processes (plasma deposition of coatings, surface treatment by laser or electron-beams, welding, surfacing, etc.) are connected with intense heat impact on a material under treatment. In this case, an abrupt temperature change of the surface occurs, which is termed as thermal shock. Various authors have discussed the role of thermal stresses in stereolithography and in selective laser sintering. A wide group of applied problems is connected with thermal shock problem to a greater or lesser extent [1-2].

The thermal shock problem was originally studied in [3] in terms of so-called theory of thermal stresses [4, 5]. This is the simplest problem formulation enabling analytical solution. Further initial problem formulation has been generalized in several directions, which was noted in reviews [6-8].

In this work, the thermal shock problem was analyzed in generalized formulation taking into account the coupling between thermal and mechanical processes. Here, the interrelation between effects is stipulated by (1) thermal conductivity, (2) dynamical (inertia) terms included in motion equations, (3) coupling between the fields strains and temperature, (4) heat inertia caused by finite rate of heat propagation and (5) dissipative effects due to plasticity. This is one of general formulations of the thermal shock problem.

2. Basic equations of the model

The problem is studied in terms of continuum mechanics. Mathematical model includes the balance equations (mass, impulse, energy):

$$\frac{\partial \rho}{\partial t} + \nabla_m (\rho u_m) = 0, \tag{1}$$

$$\frac{\partial (\rho u_k)}{\partial t} + \nabla_m (\rho u_k u_m - \sigma_{km}) = 0, \quad k = 1,2,3, \tag{2}$$

$$\rho c_\varepsilon \left(\frac{\partial T}{\partial t} + u_m \nabla_m T \right) = -\nabla_m q_m - 3K\alpha T \nabla_m u_m + s_{km} e_{km}^p, \tag{3}$$

and constitutive equations for heat flux vector:

$$q_k + \tau_q\left(\frac{\partial q_k}{\partial t} + u_m \nabla_m q_k\right) = -\lambda \nabla_k T, \quad k = 1,2,3, \tag{4}$$

and stress tensor

$$\frac{\partial s_{ik}}{\partial t} + u_m \nabla_m s_{ik} - \omega_{im} s_{km} - \omega_{km} s_{im} + \Lambda s_{ik} = 2\mu e_{ik}, \quad k = 1,2,3, \tag{5}$$

$$p = p(\rho, T), \tag{6}$$

where

$$\varepsilon_{ik} = \varepsilon_{ik}^e + \varepsilon_{ik}^p, \quad \varepsilon_{ik} = \tfrac{1}{2}\left(\nabla_i u_k + \nabla_k u_i\right), \quad e_{ik} = \varepsilon_{ik} - \tfrac{1}{3}\left(\nabla_m u_m\right)\delta_{ik},$$

$$\omega_{ik} = \tfrac{1}{2}\left(\nabla_k u_i - \nabla_i u_k\right), \quad \sigma_{ik} = -p\delta_{ik} + s_{ik}, \quad p = -\tfrac{1}{3}\sigma_{km}\delta_{km},$$

$$s_{km}s_{km} \le \tfrac{2}{3}\sigma_Y^2, \quad \Lambda = \frac{3\mu s_{km} e_{km}^p}{\sigma_Y^2} H\left(s_{km}s_{km} - \tfrac{2}{3}\sigma_Y^2\right) H\left(s_{km} e_{km}^p\right).$$

Here ρ is density, $\sigma_{ik}, \varepsilon_{ik}, \delta_{ik}$ are components of the stress tensor, strain rate tensor and identity tensor, respectively, u_k are the components of velocity vector, T is temperature, c_ε is the thermal capacity at constant strains, α is linear thermal expansion coefficient, λ is thermal conductivity coefficient, K is isothermal bulk module, τ_q is the relaxation time for heat flux, μ is shear module, σ_Y is dynamic yield point, H is Heaviside function.

Individual terms in the right part of the equation (3) describe the dissipation due to thermal conductivity ($\nabla_m q_m$), coupling between thermal and deformation processes ($3KT\alpha_T \nabla_k u_k$) and irreversible effects caused by plastic flow ($s_{km} e_{km}^p$). Here, it is supposed that stress work with plastic strains disperses completely turning into heat.

The structure of relation (4), which is usually called as the generalized Fourier's law, is identical to that of the Maxwell relaxation equation. This equation can be obtained formally by adding term $\tau_q dq/dt$ to the classical Fourier's law of thermal conductivity $q_k = -\lambda \nabla_k T$ which accounts for heat inertia. From the physical point of view, the difference between (4) and classical Fourier's law is that the generalized law takes into account a *delay effect* conditioned by the relaxation of the system to the thermodynamic equilibrium. Equation (4) can be interpreted as the generalization of Onsager relations for the case of disturbed local thermodynamic equilibrium in the fields of large (infinite) gradients of temperature field under a thermal shock. From the other hand, it can be construed as a particular case of the state equation with thermal memory [9].

Equation (5) connecting the deviators of stress tensor s_{ik} and strain rate tensor e_{ik} can be obtained with the assumption of the hypoelastic deformation law for elastic component e_{ik}^e in the additive decomposition $e_{ik} = e_{ik}^e + e_{ik}^p$ (in the hypoelastic deformation law, the Jaumann derivative is assumed as the objective measure for the rate of the stress field change), and with the assumption of the associated flow rule with von Mises plasticity condition for the plastic component e_{ik}^p.

Equation (6) is the thermal state equation for spherical stress tensor (p is hydrostatic pressure).

3. Problem formulation

Let us examine the elastic plastic half-space $x \equiv x_1 \ge 0$, surface temperature of which grows instantly from T_0 to T_b at initial time moment $t = 0$ and then stays constant, i.e.

$$T\big|_{x=0} = T_b H(t) \tag{7}$$

The present work studies the initial wave stage of half-space motion initiated by external heat action (7). Here, we analyze two types of mechanical boundary conditions: the plane $x = 0$ is free of stresses (problem I)

$$\sigma_{11}\big|_{x=0} = \sigma_{12}\big|_{x=0} = \sigma_{13}\big|_{x=0} = 0, \tag{8}$$

and the plane $x = 0$ is rigidly fixed (problem II)

$$u_1\big|_{x=0} = u_2\big|_{x=0} = u_3\big|_{x=0} = 0 \tag{9}$$

Initial conditions

$$\sigma_{ik}\big|_{t=0} = \varepsilon_{ik}\big|_{t=0} = u_k\big|_{t=0} = 0,\quad T\big|_{t=0} = T_0,\quad \rho\big|_{t=0} = \rho_0 \tag{10}$$

correspond to not stressed and not deformed quiescent half-space with uniform initial temperature and density.

Due to the uniformity of initial and boundary conditions, both problems are one-dimensional; in the half-space, a one-axis strain state is realized; hence, the following assumptions should be made for system of equations (1)–(5):

$$u_2 = u_3 = 0,\quad q_2 = q_3 = 0,\quad \varepsilon_{11} = \frac{\partial u_1}{\partial x},\quad \varepsilon_{22} = \varepsilon_{33} = 0,\quad \varepsilon_{ik} = 0,\ i \neq k,$$

$$e_{11} = \frac{2}{3}\frac{\partial u_1}{\partial x},\quad e_{22} = e_{33} = -\frac{1}{3}\frac{\partial u_1}{\partial x},\quad \sigma_{ik} = 0,\ i \neq k,\quad s_{22} = s_{33} = -\frac{1}{2}s_{11}.$$

4. Calculation results

We assume that material is steel with the following properties:

$$c_\varepsilon = 498\ \text{J/(kg}\cdot\text{K)},\quad K = 162.1\ \text{GPa},\quad \alpha_T = 10^{-6}\ \text{K}^{-1},\quad \lambda = 80\ \text{W/(m}\cdot\text{K)},$$

$$\mu = 90.6\ \text{GPa.},\quad \tau_q = 10^{-11}\text{s},\quad \sigma_Y = 0.75\ \text{GPa},\quad \rho_0 = 7.85\ \text{g/cm}^3.$$

Initial and boundary temperature are $T_0 = 293$ K and $T_b = T_0 + 1000$ K, respectively. Numerical results were obtained using the difference scheme of Godunov S.K. [10]. It is a uniform monotonic scheme with first-order accuracy; its inherent properties are similar to known gas-dynamic Godunov's scheme [11].

The results are shown in Figures 1–6. In all figures, the profiles of physical parameters corresponding to problem I are shown with thick (blue) lines; and profiles corresponding to problem II are shown with thin (dark) lines. The numbers near the curves correspond to different moments of dimensionless time $t' = t/\tau_q$, where the scale is determined by the relaxation time of heat flux. Additionally, the dotted lines in Figure 4 give the temperature for the solution of parabolic thermal conductivity equation

$$T(x,t) = (T_b - T_0)\,erfc\left(\frac{x}{2\sqrt{at}}\right),\qquad a = \frac{\lambda}{\rho c_\varepsilon}, \tag{11}$$

and dot-and-dash lines illustrate the solution of hyperbolic thermal conductivity equation [9]:

$$T(t,x)=T_0+H\left(t-\frac{x}{a_T^2}\right)\left\{T_b\,exp\left(-\frac{x}{2a_T\tau_q}\right)+\right.$$
$$\left.+\frac{x}{2a_T\tau_q}\int_0^x\frac{T_b}{\sqrt{(t-\eta)^2-x^2a_T^{-2}}}I_1\left(\frac{\sqrt{(t-\eta)^2-x^2a_T^{-2}}}{2\tau_q}\right)exp\left(-\frac{t-\eta}{2\tau_q}\right)d\eta\right\}, \tag{12}$$

where $a_T=\sqrt{a\tau_q^{-1}}$, I_1 is a modified Bessel function of first kind, H is the Heaviside function.

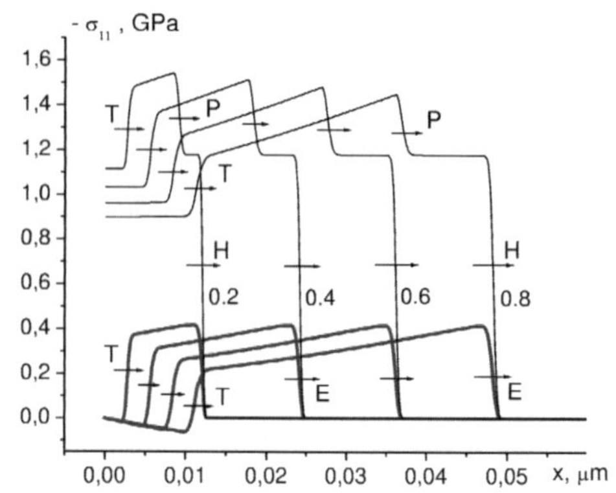

a

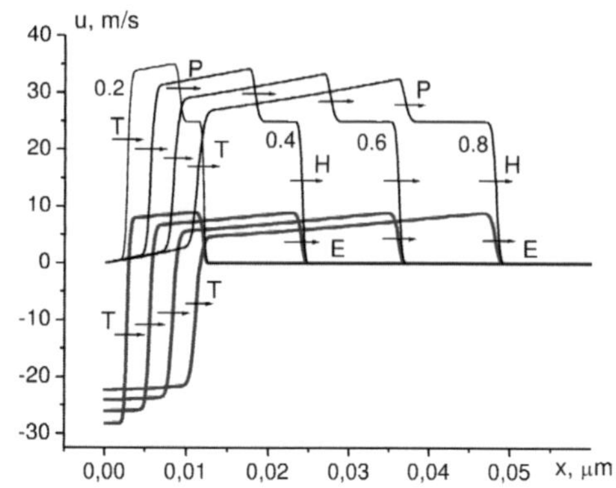

a

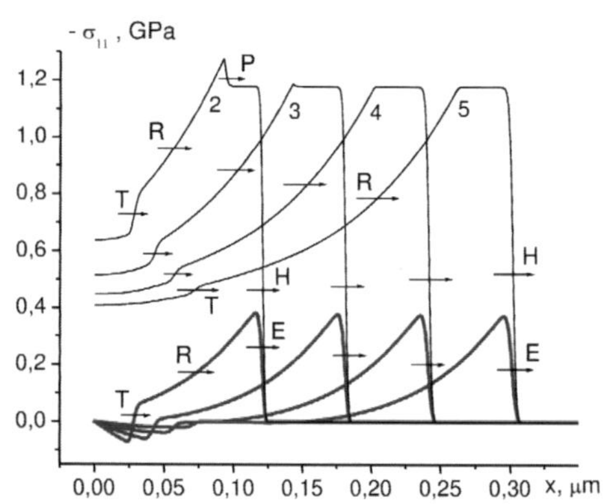

b

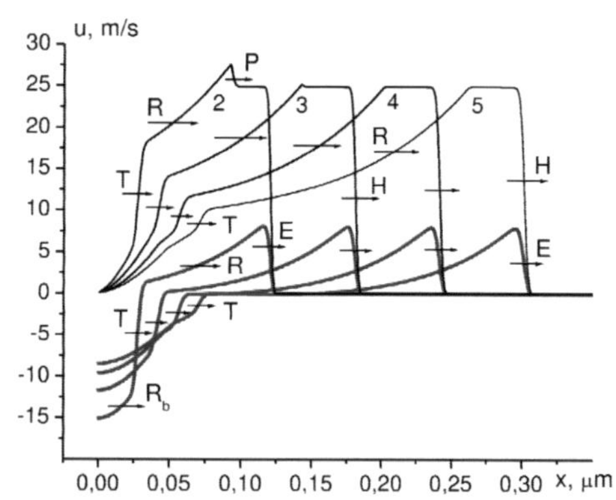

b

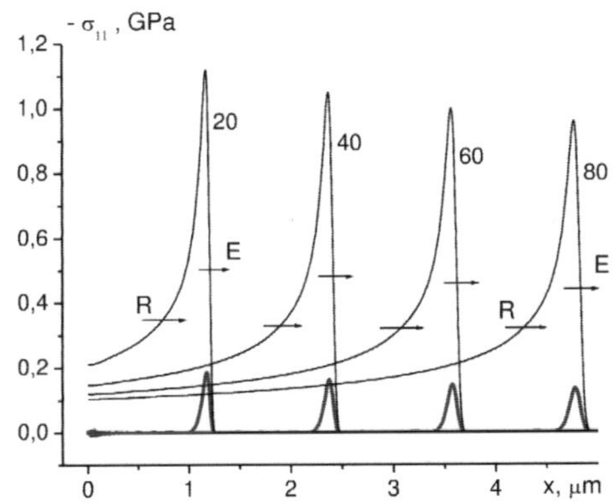

c

Figure 1. Stress profiles for sequential time moments. Time intervals: a) 0–1 s; b) 1–10 s; c) 10–100 s.

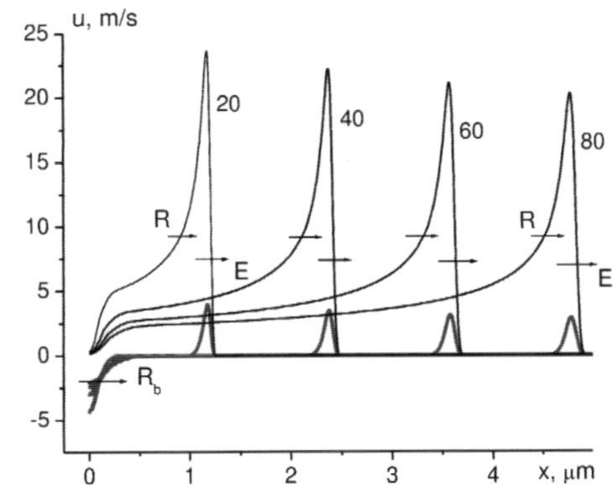

c

Figure 2. Velocity profiles for sequential time moments. Time intervals: a) 0–1 s; b) 1–10 s; c) 10–100 s.

Let us consider the qualitative features of the wave picture of motion developing at the initial stage of thermal shock.

First of all, we should note that there are several different velocities for the propagation of thermomechanical perturbations. Taking into account the coupling of heat and deformation processes, the propagation rates for quick a_1 and slow a_2 waves can be presented as [10]:

$$a_{1,2} = \sqrt{\frac{1}{2}\left(a_E^2 + a_T^2\right) \pm \frac{1}{2}\sqrt{\left(a_E^2 - a_T^2\right)^2 + 4T\left(3K\alpha a_T\right)^2 c_\varepsilon^{-1}\rho^{-2}}}\,, \tag{13}$$

where the value of $a_E = \sqrt{(K_S + 4\mu/3)\rho^{-1}}$ corresponds to the propagation velocity of longitudinal elastic waves, and $a_T = \sqrt{a/\tau_q}$ is the thermal wave velocity for the case when coupling effects are omitted. Here K_S is adiabatic bulk module, $a = \lambda(\rho c_\varepsilon)^{-1}$ is thermal diffusivity coefficient. In addition, there are still two velocities, $a_H \approx a_E$ and $a_P \approx \sqrt{K_S/\rho}$; a_H correspond to the velocity of elastic predecessor, a_P is the plastic wave velocity.

As a rule, the coupling coefficient for metals is small [4, 9]. When we neglect this effect and remove the last term from formula (13), the approximate equalities can be written as

$$a_1 \approx a_E \approx a_H, \qquad a_2 \approx a_T.$$

In the case under study, $a_H \approx a_E = 6022$ m/s, $a_P = 4569$ m/s, $a_T = 1431$ m/s.

According to above, elastic-plastic wave initiated by thermal shock, has several specific parts. For convenience, we will term the waves propagating with velocities a_E, a_H, a_P, a_T as E-, H-, P-, and T-waves, respectively. Figures. 1–4 show that the state of medium alters in these waves unevenly. The sharp growth in temperature and pressure occurs in the front of T-wave. Behind this wave, the intense expansion of the substance is observed that leads to the formation of the impulse of tensile stresses in problem I (Figure 1), and the motion velocity of free surface in the negative direction of x-axis is tens meters per second (Figure 2). The processes developing in the T-wave has fierce irreversible character which is obvious from the pressure profiles (Figure 3): first, "over-compressing" occurs, than "over-expansion" of the substance emerges, and only at $t \sim 10\tau_q$ the pressure limits to a reversible value (p = 0.5 GPa) corresponding to the regime determined by boundary conditions ($\sigma_{11} \equiv s_{11} - p = 0$).

The region of continuous flow lies between E- and T-waves; it corresponds to the unloading wave, marked in the figures by letter R. With time, this wave decreases the stress to zero; further, T- and E-waves look as separated in problem I (Figures. 1–3, $t' = 20-80$). However, the interrelation between them persists, and the processes developing in T-wave affect appreciably E-wave attenuation.

Quantitative difference in the solutions of problems I and II, shown clearly in Figures. 1–3, are evidently connected with the condition of hindered deformation in problem II. The material of the half-space having endured a large volume expansion behind T-wave, cannot expand freely in the negative direction of x-axis due to the fixation condition. As a result, the pressure behind T-wave in problem II is higher than that in problem I, and stress σ_{11} stays compressive. In addition, the adjugate P-jump appears in problem I. It is connected with the intensity of tangential stresses behind H-wave reaching maximum possible value corresponding to the von Mises yield condition. Stress σ_{11} in H-wave equals to Hugoniot elastic limit on shock adiabat. The adiabat has a kink in this point, and the stress exceeding the Hugoniot limit propagates with the volume wave velocity a_P.

Figure 4 shows that for small times, the numerical solution agrees well with the analytical solution of hyperbolic thermal conductivity equation (12), and for large times, with the solution of parabolic equation (11). For any time, the contribution of mechanical effects to temperature field can be ignored. Negligibly small effect of elastic plastic strains in the temperature change is shown in Figure 5. Elastic wave of limiting intensity, H-wave, increases the temperature for 1.5 degrees.

The stress profiles $\sigma_{11}(x)$ are presented in Figure 6 for three time moments $t' = 0.1, 0.2, 0.3$ seconds (problem I).

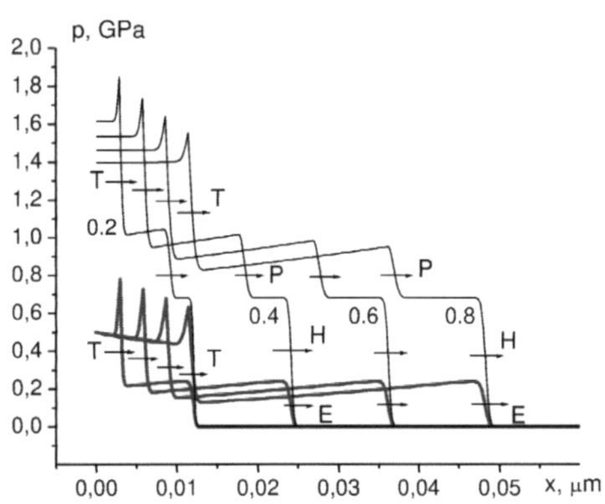

a

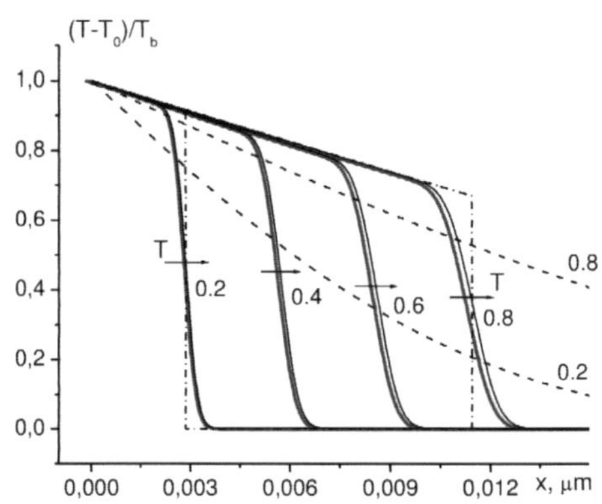

a

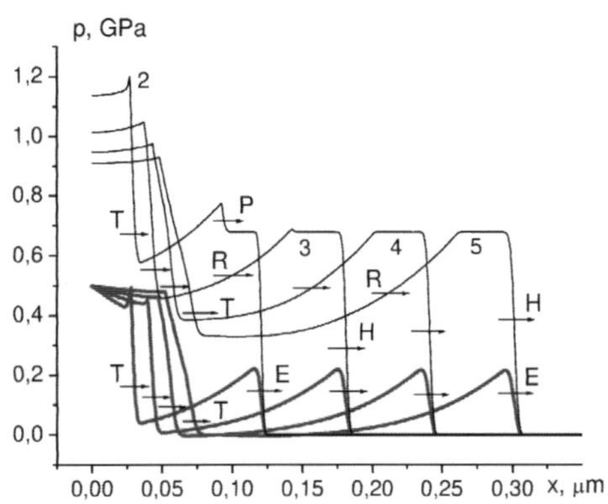

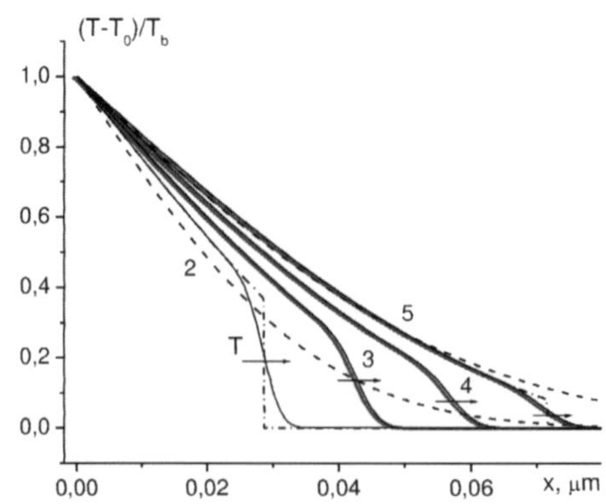

IOP Conf. Series: Materials Science and Engineering **140** (2016) 012022 doi:10.1088/1757-899X/140/1/012022

b

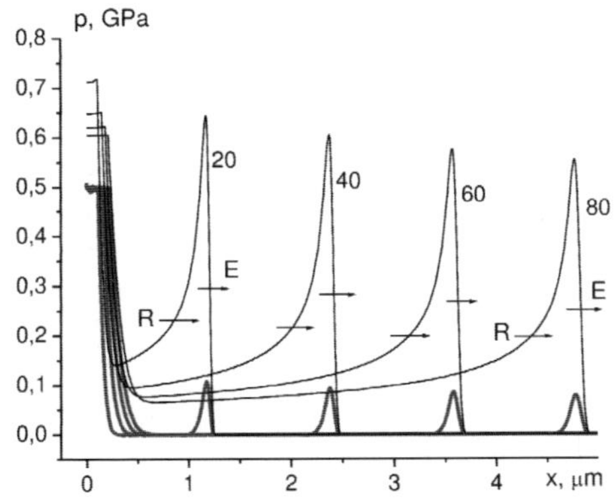

c

Figure 3. Pressure profiles for sequential time moments. Time intervals: a) 0–1 s; b) 1–10 s; c) 10–100 s.

b

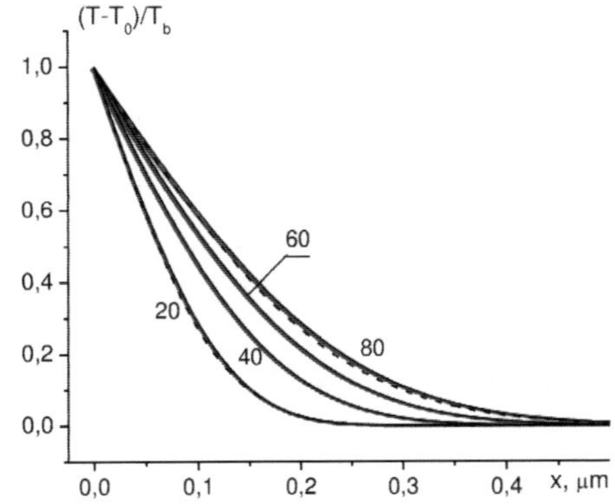

c

Figure 4. Temperature profiles for sequential time moments. Time intervals: a) 0–1 s; b) 1–10 s; c) 10–100 s.

The known thermoelastic solution [12]

$$\sigma_{11}(x,t) = -A[D(x,t) + E(x,t)], \tag{14}$$

where

$$A = 3K_T\alpha_T(T_b - T_0), \quad E(x,t) = \frac{1}{2}exp\left[\frac{a_E^2}{a}\left(t - \frac{x}{a_E}\right)\right]H\left(t - \frac{x}{a_E}\right)$$

$$D(x,t) = \frac{1}{2}exp\left[\frac{a_E^2}{a}\left(t + \frac{x}{a_E}\right)\right]erfc\left(\frac{x}{2\sqrt{at}} + a_E\sqrt{\frac{t}{a}}\right) + \frac{1}{2}exp\left[\frac{a_E^2}{a}\left(t - \frac{x}{a_E}\right)\right]erfc\left(\frac{x}{2\sqrt{at}} - a_E\sqrt{\frac{t}{a}}\right),$$

is also shown here for the same times for the sake of comparison.

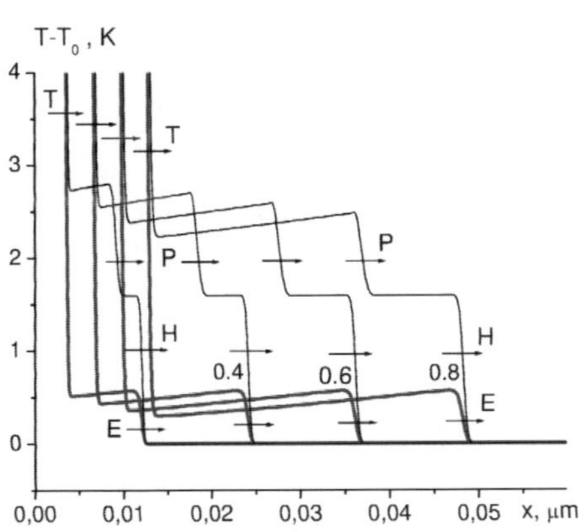

Figure 5. Temperature profiles for sequential time moments

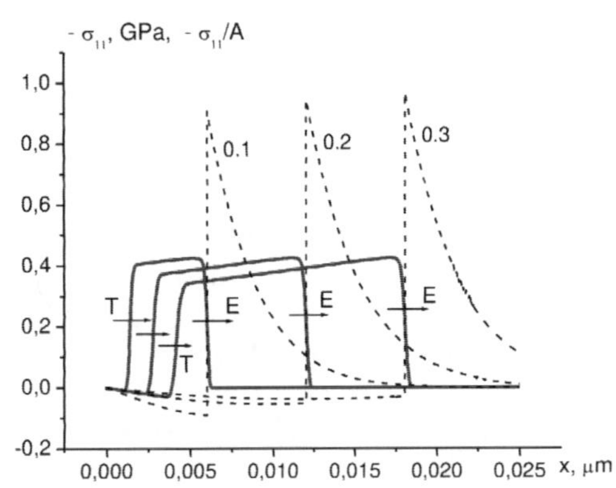

Figure 6. Stress profiles for sequential time moments. Comparison with classical thermoelastic solution (dotted line).

Diffusion part of the problem described by function $D(x,t)$ corresponds to infinite velocity of heat propagation; hence, compressive stresses appear immediately in each point of the half-space. The second term $E(x,t)$ determines the elastic wave $E(x,t)$ propagating with velocity a_E. Stress σ_{11} changes unevenly by value A in the front of this wave (in the example under study, $A = 4.86$ GPa) and turns tensile. The solution described by equation (14) does not even qualitatively conform with the solution obtained in this work. Quantitative comparison is useless because the solution (14) is correct, if the condition $(T - T_0)T_0^{-1} << 1$ was satisfied, which is not in considered case.

5. Conclusion

The problem of thermal shock is studied here for the half-space with the properties described by an elastic plastic model taking into account dynamical effects, heat inertia, coupling between heat and mechanical fields. The problem was solved numerically using S.K. Godunov's method.

The obtained results show that wave picture accompanying a thermal shock is incomparably richer and more complex, as compared to those predicted by known thermoelastic solutions.

In this connection, we note that the evaluations of the thermal strength of materials based on the solutions of type (14) used in the literature can be inconsistent, because the fracture of surface layers, if any, develops at the initial (wave) stage of thermal shock, when the relative role of the factors connected with mechanical and heat inertia is maximal and actually determines the following evolution of a thermal stresses field. Therefore, residual stresses under local thermal action do not correspond to the solutions of quasistatic problems [13].

Results obtained for the heat part of the problem once more corroborate the known situation about weak influence of the coupling effect on temperature distribution. On all occasions, this is correct for typical construction materials and alloys on the basis of iron and aluminum. For these materials, one can neglect the contribution of mechanical processes to temperature field and use the hyperbolic equation (for $t \leq 10\tau_q$) or more simple parabolic thermal conductivity equation (for $t \geq 100\tau_q$). However, the relative contribution of plastic strains for nonmetallic materials can appreciably different in comparison with the example studied here.

In conclusion we should note that the thermomechanical wave attenuation is strongly affected by all the factors named in the initial part of the paper (dynamic effects, heat inertia, coupling) and by boundary conditions determined for the free surface of a half-space.

Acknowledgments

This work was supported by the Ministry of Science and Education of the RF, project No. 11.815.2014/K.

References

[1] Rykalin N N, Zuev A A and Uglov A A 1978 *The foundation of electron-beam treatment of materials (in Russian)* (Moscow: Mashinostroenie) p 239
[2] Lawrence E Murr, Sara M Gaytan, Diana A Ramirezb, Edwin Martinez, Jennifer Hernandez, Krista N Amato, Patrick W Shindo, Francisco R Medina and Ryan B 2012 Wicker Metal Fabrication by Additive Manufacturing Using Laser and Electron Beam Melting Technologies *J. Mat. Sc. & Tech.* **28** **1** 1–14
[3] Danilovskaya V I 1950 Thermal Stresses in an Elastic Half-Space Arising After a Sudden Heating of Its Boundary (in Russian) *Prikl. Mat. Mech.* **14** **3** 316-318
[4] Boley B A and Weiner I H 1960 *Theory of thermal stresses* (New York: Wiley)
[5] Parkus H 1959 *Instutionäre Wärmespannungen* (Vienna: Springer)
[6] Kartashev E M, Parton V Z 1991 *Dynamic thermoelasticity and problems of thermal shock (in Russian)* (Itogi Nauki I techniki VINITI. Ser. "Mechanics of deformable solid body") **22** 55-127
[7] Xiangmin Zhou 2008 *Journal of Thermal Stresses* **31** **7** 663-664

[8] Richard B Hetnarski and Eslami M. Reza 2009 *Thermal Stresses – Advanced Theory and Applications* (Berlin: Springer)
[9] Lykov A V 1978 *Heat and mass transfer. Reference book (in Russian)* (Moscow: Energiya)
[10] Demidov V N 1999 Godunov's Method for generalized one-dimensional dynamical problems of thermal elastic plasticity – *in "Mathematical modelling in synergetic systems". All-Russian scientific conference (Tomsk-Ulan-Ude July, 20-23, 1999)* pp 226-228
[11] Godunov S K 1958 Difference method for the calculation of discontinuous solutions in hydrodynamics (in Russian) *Mathematical collection* **47** **3** 271-306.
[12] Nowacki V 1970 *Dynamiczne zagadnienia termosprezystosci. Warshava, (Nowatskii V. Dynamical problems of thermal elasticity (translated to Russian)* (Moscow: Mir) p 250
[13] Burenin1 A A, Dats E P and Murashkin E V 2014 Formation of the Residual Stress Field under Local Thermal Actions *Mechanics of Solids* **49** **2** 218–224

IOP Conf. Series: Materials Science and Engineering **140** (2016) 012023 doi:10.1088/1757-899X/140/1/012023

Fabrication of porous silicon nitride ceramics using binder jetting technology

L Rabinskiy[1] , **A Ripetsky**[1], **S Sitnikov**[1], **Y Solyaev**[1,2], **R Kahramanov**[1]

[1] Moscow Aviation Institute, Volokolamskoe av., 4, Moscow, Russia
[2] Institute of Applied Mechanics Russian Academy of Science, Leningradskiy av., 7, Moscow, Russia

E-mail: f9_dec@mai.ru

Abstract. This paper presents the results of the binder jetting technology application for the processing of the Si_3N_4-based ceramics. The difference of the developed technology from analogues used for additive manufacturing of silicon nitride ceramics is a method of the separate deposition of the mineral powder and binder without direct injection of suspensions/slurries. It is assumed that such approach allows reducing the technology complexity and simplifying the process of the feedstock preparation, including the simplification of the composite materials production. The binders based on methyl ester of acrylic acid with polyurethane and modified starch were studied. At this stage of the investigations, the technology of green body's fabrication is implemented using a standard HP cartridge mounted on the robotic arm. For the coordinated operation of the cartridge and robot the specially developed software was used. Obtained green bodies of silicon powder were used to produce the ceramic samples via reaction sintering. The results of study of ceramics samples microstructure and composition are presented. Sintered ceramics are characterized by fibrous α-Si_3N_4 structure and porosity up to 70%.

1. Introduction

Silicon nitride ceramics have excellent mechanical, thermal and dielectric properties. Silicon nitride parts are widely used in different technical fields [1, 2], particularly in jet engine structural components, in the fuel systems, in the braking system components. Porous silicon nitride ceramics are applied in cooling system, filtration components and as refractory materials.

Over the last twenty years, several additive fabrication technologies for silicon green bodies and silicon nitride parts, which possess a complex geometry and different pore volume, were suggested [3-11]. The laminated object manufacturing (LOM) process of silicon nitride products was suggested in the work [3]. The laying of thin polymer film, which contains particles of α-Si_3N_4, was carried out. Layers of tape were stacked together with a preheated roller and cut off with a laser to fabricate a defined shape product. Up to 90% β-Si_3N_4 and 6% porosity were formed during the material sintering process. The flexural strength of the material was up to 900 MPa. Fabrication of micro-sized silicon nitride nozzles using mould shape deposition manufacturing technology was applied during the work [4]. Fused deposition modelling was used in the work [5]. Fusing layers of the product were carried out by heating and extruding of thermoplastic polymer filament with 55% silicon nitride powder content. The technology requires careful control over the powder dispersion – the agglomerate size should not exceed the nozzle diameter (250 µm). Various schemes of filament laying were used to produce isotropic material. The flexural strength was about 900 MPa, and the fracture toughness of 8.5 MPa·m^{0,5}.

The papers by Cappi et al. [6, 7] introduced the fabrication of silicon nitride ceramics by direct inkjet printing technology using HP DeskJet 930c inkjet printer. Printing was carried out using suspensions with volume content of about 30% silicon nitride, 0.7 μm average size, and additives of yttrium aluminum garnet and boron nitride. The slurry was formed by the mixture of polyacrylic and carboxylic acids with the addition of ethanol and ethylene glycol. The printer used two cartridges that separately applied the main and supporting materials by suspending silicon nitride particles. The composition of final samples includes β-Si3N4, silicon oxynitride, pure silicon and yttrium silicate. The material fracture toughness was 4.4 MPa·m^{0,5}, Vickers hardness – 17 GPa.

In the recent papers [8-10] it was considered three-dimensional printing technology for highly porous silicon nitride ceramics containing more than 70% pores. The powder containing more than 90 wt.% of α-Si_3N_4 was used as a feedstock. Products were fabricated by three-dimensional printing carried out on a standard 3D-printer using water-based solvents. It was found that obtained material has the highest porosity (68% versus 45% for traditional technology), and its properties are relatively low. Due to high porosity, its strength, elastic moduli, fracture toughness and hardness were significantly reduced. The capability of good dielectric properties of the material was mentioned. The work [10] presented the development of technology to produce porous silicon nitride composites reinforced with in-situ formed twinned silicon carbide nanowires. An increase of SiC content in the fabricated composites resulted in the growth of permittivity. The reinforced SiC samples had a minimal reflection coefficient at microwave frequencies and had excellent absorption properties.

In the recent paper [11] small-size complex shape samples were fabricated using LOM technology. The films were prefabricated using suspensions of silicon nitride powder with averaged particles size of 130 nm. These pieces were stacked, pressing and then sintered at temperature of 1800 °C. The porosity of the obtained material was 50% and the flexural strength was 475 MPa.

It could be mentioned that one of disadvantages of existing silicon nitride additive manufacturing technologies is a difficulty of feedstock preparation: powder dispersion in the direct inkjet printing or preparation of films with Si_3N_4 particles in the LOM technology. FDM technology requires solving the additional problem of thermoplastic binder removal from the fabricated products. This is a time-consuming and non-environmentally friendly process, which often results in defects of the parts with complex geometry. One of the technologies with no such manufacturing operations is a binder jetting. This technology is widely used for the synthesis of gypsum, mortar sand etc.

In the present work, we offer the method of silicon green bodies fabrication using binder jetting technology. The silicon nitride ceramics are obtained using the pressureless reaction sintering process. This technology does not use the mentioned complex preliminary processes. In this case, the binder is injected into each applied layer of dry ceramic powder with controlled thickness. Layer-by-layer bonding of the powder particles leads to the shaping of the future products or green bodies. Part of the powder in each layer that not exposed to a binder, acts as a support for the synthesized parts with complex shapes. General description and principles of the binder jetting technology are presented, for example, in the work [12]. In order to implement the three-dimensional printing process we used a modified system based on an industrial robotic arm and printing block of original design. Software for printing process control was developed. The study and comparison of two variants of binders and silicon powder preparation methods were undertaken. The composition and microstructure of the first obtained ceramic samples, which have high porosity, were studied.

2. Experimental Procedure

2.1. Powders

In the experiments the commercially available technical silicon powder KR00 (TU-1711-5072130-01-2012) with fraction of 10-100 mm was subjected to vibrating grinding in ball mills SVM-2. The duration of grinding was varied in the range of 9-36 hours, grinding bodies were the high-strength steel balls with diameter of 16 mm. Grinding was carried out in the presence of surfactants: oleophilic (oleic acid 0.6 wt.%) or hydrophilic (sulfanol 0.6 wt.%). The average diameter of powder particles was 0.4-1.5 μm. This value of particles ensures the successful and complete process of reaction sintering in the temperature range of 1170-1500 °C. The special criterion for the selection of the required fineness of the powder was the minimal weight content of the selected binder required to obtain the reliable formation of silicon green bodies. The particle size was controlled by a combined electro-acoustic analyzer Dispersion Technology DT 1201.

It should be noted that coarse powder requires less binder but does not provide the complete reaction sintering process. Using a large amount of the binder for the ultrafine powder tends to reduce the accuracy and speed of modeling, elongation of the drying process time and increase the number of defective articles. Therefore, two compromise variants of the silicon powder preparation were developed to solve mentioned mutually exclusive issues. In the first case, silicon powder was coated by the aluminum isopropoxide to decrease the wettability of a binders' solution on solid particles surfaces. After hydrolysis and drying, the nanoscale layer of aluminum oxide was formed on the silicon particles. Size of silicon particles was insignificantly increased but the amount of binder used in the synthesis was reduced. In the second case, the ultrafine silicon powder pre-aggregated (using binder that employed during synthesis) to the state of granules of 50-100 μm, which also reduced the amount of binder used in the synthesis.

2.2. Binders

Two types of binders were used to prepare the green-body preforms:

- A first binder based on a copolymer of methyl ester of acrylic acid with polyurethane with additives of photoinitiator and a diluent – the monomer acrylic acid (the binder 1). Solidification of this binder was initiated by the ultraviolet radiation in the process of three-dimensional printing and proceeded within 20-100 sec depending on thickness of the curable layer. In this technique, the silicon powder coated with alumina is used.
- The binder based on polysaccharide – modified starch (binder 2). The pre-aggregation of silicon powder using this binder was conducted. Consolidation of the powder granules was performed by applying the aqueous solution through the cartridge nozzle during the printing process.

2.3. Binder jetting procedure

Binder jetting procedure is based on the layer-by-layer binder impregnation on the silicon powder in the area of cross-section of the produced sample. To implement this process, the robot arm KUKA KR 6 (KUKA AG) was modified. On the flange of the robot it was installed the printing unit, which consists of Arduino controller, InkShield controller and HP 6022B cartridge with integrated nozzle array. Resolution 96 ppi of the cartridge ensures the efficient printing on media with complex structure. The cartridge is controlled by Arduino system through the combinations of control signals for the nozzle array. For this, the special program was developed using C programming language. Data transmission from the PC to the printing unit was done through a serial COM-port. The developed software provides synchronization of the controller of the printing block with a robotic arm. The swivel rotary table is used as the base for the samples in three-dimensional printing process. The powder supply system on the work platform is carried out using roller mechanism, which provide full and uniform laying of each layer of the powder in the chamber. Once a layer of the powder is supplied, the inkjet printheads print the lower section of the part on the smooth surface of the powder, gluing particles together. Thereafter, the plunger lowered the working platform onto the layer thickness down and a new layer of the powder is applied. The printing unit prints the next section on the new layer surface, gluing it to the previous layer. This process is repeated for all sections of the sample. The technology does not require solid or attached supports during the printing process, so all amount of unused powder can be used again.

2.4. Reaction sintering
Reaction sintering of ceramic samples was carried out in the vacuum furnace, and consisted of the following main operations:

- Vacuumization and heating up to 360 °C at the rate not exceeding 65°C per hour for degassing;
- Heating up to 1170 °C at a rate not exceeding 65 °C per hour to warm the samples up to the initiation temperature of the reaction of nitrogen supplied to the furnace with silicon particles in the printed green bodies;
- Further heating up to the 1400 °C at a rate not exceeding 20 °C per hour, which allows the complete nitridation of the samples during for 8-10 hours;
- Cooling of the samples with the furnace for at least 1 hour.

2.5. Binder jetting procedure
Density of materials were determined by direct measurement. X-ray diffraction analyses have been performed using ALRX'TRA (Thermo-FisherScientific) diffractometer with copper anode using the Crystallographica Search-Match Version 3.1.0.0 software and PDF-2 crystallographic database. The study of microstructure of the samples fracture surfaces and X-ray microanalyses were conducted using two-beam scanning electron microscope Versa 3D LoVac. The surface of the samples was coated with gold.

3. Results and Discussion
The weight content of the binder which is necessary for reliable forming of the green bodies during three-dimensional printing via binder jetting using different powders is presented in Table 1. Analysis of the results shows that used surfactants during the grinding, are not effective for the binder 1 and it requires additional studies. Therefore, further experiments with this binder 1 were carried out with the silicon powder obtained by grinding in a vibrating mill without surfactant for 18 hours. The second binder showed an affinity for sulfanol. Therefore, experiments with binder 2 were carried out with silicon powder obtained by milling for 9 hours in the presence of sulfanol.

Table 1. The weight contents of the binders, which are necessary for reliable forming of the green bodies during three-dimensional printing via binder jet using different fractions of silicon powders.

Milling time, h	Surfactant	Particles size, μm	Binder 1	Binder 2
9	oleic acid	0.45	do not mix	50
18	oleic acid	0.3	do not mix	70
36	oleic acid	0.1	do not mix	100
9	sulfanol	0.6	55	12
18	sulfanol	0.4	77	18
36	sulfanol	0.15	105	24
9	-	3	32	20
18	-	1.5	45	32
36	-	0.5	60	40

As result of X-ray analysis of samples subjected to reaction sintering, it was found that the resulting ceramics consists of α-Si_3N_4 with possible content of silicon oxynitride up to 10%. The resulting X-ray diffraction pattern and its comparison with the standard data for the alpha-Si_3N_4 is shown on Figure 1. As result of X-ray microanalysis, the elemental composition of the samples fracture surfaces was determined which is listed in the Table 2. It was found that usage of the binder 1 tend to increase the carbon (presumably SiC) content in the ceramics composition. The presence of aluminum in the ceramic composition is explained by the preliminary treatment of silicon powder with aluminum isopropoxide.

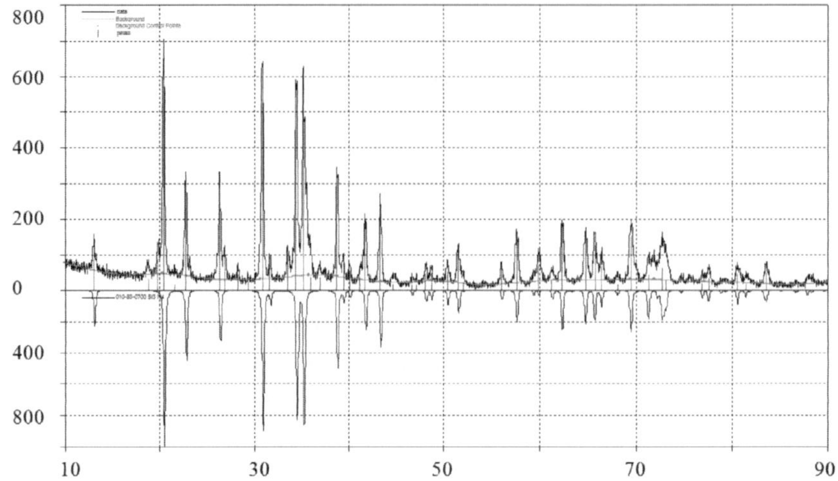

Figure 1. Results of X-ray analysis of ceramic samples composition (top pattern) and the standard data for the alpha-Si_3N_4 from the PDF-2 database (bottom pattern).

Table 2. The weight content (%) of the chemical elements identified by X-ray microanalysis on the fracture surfaces of ceramic samples prepared using different binders.

Elements	Binder 1	Binder 2
Si	68.3	58.6
N	14.5	26.4
C	11.1	7.4
O	3.5	6.6
Al	2.6	1

Investigation of the microstructure of the obtained ceramic samples is shown in Figures 2 and 3. Samples have a fibrous structure with dispersed inclusions (ceramic grains) with size of 1-20 μm. The first variant of the binder allows to obtain a denser structure consisting of shorter and thicker fibers, compared with the second variant of the binder. Porosity of the samples obtained using the binder of the first and second type was up to 60% and 70%, respectively.

a 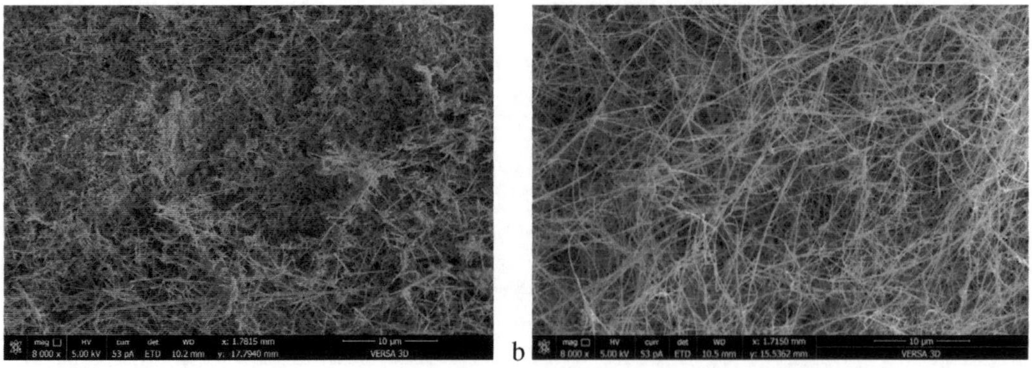 b

Figure 2. Microstructure of ceramic samples obtained using binder jetting and reaction sintering technologies using binder 1 (a), binder 2 (b).

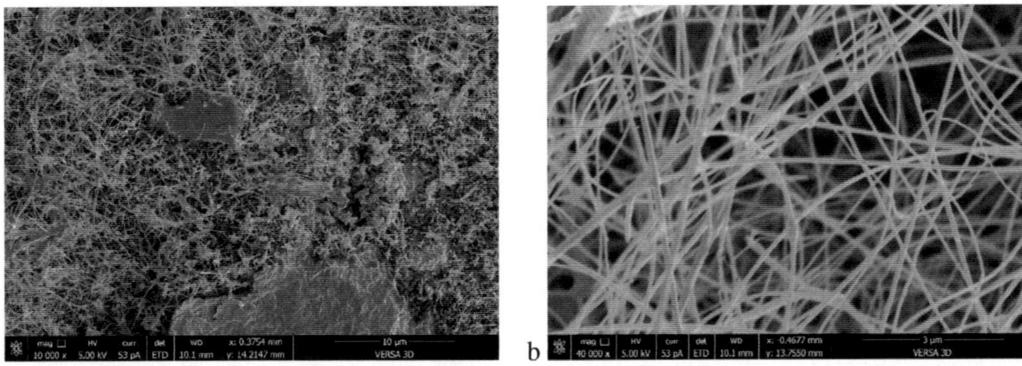

Figure 3. Microstructure of ceramics obtained using binder 2: large inclusions (a) and high-resolution images of α-Si_3N_4 fibers (b).

4. Conclusions

The development of silicon nitride ceramic additive fabrication using binder jetting and reaction sintering technology was carried out. The control system of binder jetting three-dimensional printing process was developed. The method of silicon powder preparation was offered. Two types of binders based on acrylic acid with polyurethane and based on modified starch were used. Optimal parameters of silicon powder preparation were determined. Porous silicon nitride samples, for which the composition and microstructure parameters were examined, were fabricated resulting from the reaction sintering of the obtained preforms. It was found that the material consists mostly of α-Si_3N_4 fibers with particle inclusions of 1-20 μm. Silicon oxynitride and silicon carbide with total volume content of not more than 20% are also present in the material composition.

Acknowledgments

The study was performed by Federal Target Program of the Ministry of Education and Science of the Russian Federation "Research and Development 2014-2020", agreement №14.577.21.0171.

References

[1] Krstic Z and Krstic V 2012 J. Mater. Sci. **47** 535
[2] Somiya S 2013 *Handbook of Advanced Ceramics: Materials, Applications, Processing, and Properties* (NY: Academic Press)
[3] Rodrigues S *et al* 2001 Proc. of the SFF Symp. 1
[4] Liu H *et al* 2004 Rapid Prototyping J. **10** 239
[5] Iyer S *et al* 2008 Int. J. Appl. Ceram. Techn. **5** 127
[6] Cappi B *et al* 2008 J. Europ. Ceram. Soc. **28** 2625
[7] Cappi B *et al* 2011 J. Amer. Ceram. Soc. **94** 111
[8] Li X *et al* 2012 Scr. Mater. **67** 380
[9] Li X *et al* 2012 Mater. Sci. Eng. A **549** 43
[10] Duan W *et al* 2015 Mater. Lett. **159** 257
[11] Liu S *et al* 2015 Mat. & Des. **66** 331
[12] Gibson D *et al* 2013 *Additive Manufacturing Technologies. 3D Printing, Rapid Prototyping, and Direct Digital Manufacturing* (NY: Springer)

Institute of Physics
Dirac House, Temple Back
Bristol BS1 6BE UK

CURRAN ASSOCIATES INC.
proceedings.com

ISSN: 1757-8981
ISBN 978-1-5108-2819-3

7th International Conference on Applied Physics and Mathematics (ICAPM 2017)

Journal of Physics: Conference Series Volume 814

Tokyo, Japan
20 – 22 January 2017

IOP
Institute of Physics